The Intersection of Change Management and Lean Six Sigma

The Intersection of Change Management and Lean Six Sigma
The Basics for Black Belts and Change Agents

Randy K. Kesterson

CRC Press
Taylor & Francis Group
Boca Raton London New York

CRC Press is an imprint of the
Taylor & Francis Group, an **informa** business

CRC Press
Taylor & Francis Group
6000 Broken Sound Parkway NW, Suite 300
Boca Raton, FL 33487-2742

© 2018 by Taylor & Francis Group, LLC
CRC Press is an imprint of Taylor & Francis Group, an Informa business

No claim to original U.S. Government works

Printed on acid-free paper

International Standard Book Number-13: 978-1-1382-1702-7 (Paperback)
International Standard Book Number-13: 978-1-138-10297-2 (Hardback)
International Standard Book Number-13: 978-1-315-11689-1 (eBook)

This book contains information obtained from authentic and highly regarded sources. Reasonable efforts have been made to publish reliable data and information, but the author and publisher cannot assume responsibility for the validity of all materials or the consequences of their use. The authors and publishers have attempted to trace the copyright holders of all material reproduced in this publication and apologize to copyright holders if permission to publish in this form has not been obtained. If any copyright material has not been acknowledged please write and let us know so we may rectify in any future reprint.

Except as permitted under U.S. Copyright Law, no part of this book may be reprinted, reproduced, transmitted, or utilized in any form by any electronic, mechanical, or other means, now known or hereafter invented, including photocopying, microfilming, and recording, or in any information storage or retrieval system, without written permission from the publishers.

For permission to photocopy or use material electronically from this work, please access www.copyright.com (http://www.copyright.com/) or contact the Copyright Clearance Center, Inc. (CCC), 222 Rosewood Drive, Danvers, MA 01923, 978-750-8400. CCC is a not-for-profit organization that provides licenses and registration for a variety of users. For organizations that have been granted a photocopy license by the CCC, a separate system of payment has been arranged.

Trademark Notice: Product or corporate names may be trademarks or registered trademarks, and are used only for identification and explanation without intent to infringe.

Visit the Taylor & Francis Web site at
http://www.taylorandfrancis.com

and the CRC Press Web site at
http://www.crcpress.com

To my wife, Susan, and my three kids: Michael, Nicole, and Chase

Contents

List of Abbreviations xi
Foreword xiii
Preface xv
Working at the Intersection xix
Introduction xxi
Thank You xxiii
About the Author xxv

SECTION I

Chapter 1 Why Should You Care? 3
              Change Receptiveness of the Organization 6
              A Catalyst for Success 8
              Respectful and Respected Leadership 8
              \kə-ˌla-bə-ˈrā-shən\ 8
              The Collaboration Trade-Off 8
              Key Takeaways 9
              Reference 9

Chapter 2 Change Is Good! 11
              Key Takeaways 13

Chapter 3 Change Is Hard! 15
              Key Takeaways 20
              Reference 20

Chapter 4 Resistance Is Real 21
              The Change Curve 21
              Key Takeaways 24

Chapter 5 Minimizing Resistance 25

The Story of the Shoe that Changed My Perspective on Change 25
Key Takeaways 30

Chapter 6 Why?: The Most Important Word in This Book 31

Why? 31
The Targets of the Change 32
People Analytics Tools 33
Key Takeaways 35
Reference 35

Chapter 7 Resistance in Your Personal Life? 37

Key Takeaways 37
Reference 38

SECTION II

Chapter 8 What Is Organizational Change Management? 41

The ADKAR Model (by Prosci, Inc.) 42
Managed Change Model (by LaMarsh Global) 42
    Resistance to Leaving the Current State 43
    Resistance to Moving to the Future State 43
    Resistance to the Change Plan 44
The Five Stages 45
Emotional Support 47
    A Chaordic Change, Discovered in an Unlikely Place 47
Friction Equation 49
Organizational Change Management 50
    Some Lessons Learned on How Can I Help? 50
        The Reinforcement Paradox? 51
Key Takeaways 52
References 53

Contents • ix

Chapter 9 Project Risk Assessment 55

    Some Changes are Bad Changes 57
    How to Prevent Bad Projects 58
    Key Takeaways 58

Chapter 10 Some Basic Organizational Change Management Tools 59

    Stakeholder Analysis Tool 59
        Stakeholder 59
        What Is the Tool Used For? 59
        How to Use the Tool? 60
        When to Use the Tool? 61
        What to Do with the Results? 61
    Risk Analysis Tool 62
        What Is the Tool Used For? 62
        How to Use the Tool? 62
        When to Use the Tool? 63
        What to Do with the Results? 63
    Communication Plan 63
        What Is the Tool Used For? 63
        How to Use the Tool? 64
        When to Use the Tool? 64
        What to Do with the Results? 64
    Levers You Can Pull 65
        Communication 66
        Incentives 66
        Training and Development 67
    Key Takeaways 68
    Reference 68

Chapter 11 When to Ask for Help 69

    OCM Intuition 70
    Key Takeaways 71

Chapter 12 The Intersection of Organizational Change Management and Lean Six Sigma 73

    A Note to the LSS Program/Deployment Manager 75
    Key Takeaways 76

Chapter 13 People *Are* Different 77

    Introduction to the WorkPlace Big Five Profile 78
    Need for Stability (N) 79
    Extraversion (E) 79
    Originality (O) 79
    Accommodation (A) 80
    Consolidation (C) 80
    Change Propensity 82
    People and Facts and Data 82
    People Analytics 83
    Key Takeaways 84
    References 85

Chapter 14 The Final Word 87

    Key Takeaways 88

SECTION III

Chapter 15 Interviews with Experts 91

    The Interview Questions 92

Chapter 16 Mini-Biographies of the Interviewees 159

SECTION IV

Appendix A: Competencies for the Successful Black Belt 165

Appendix B: Coaching and Mentoring for the Black Belt 167

Appendix C: The Shingo Principles 169

Appendix D: A Brief History of Continuous Improvement 171

Appendix E: Where to Go for More Information 179

Appendix F: Acknowledgments 181

Bibliography 183

Index 185

List of Abbreviations

ADKAR	awareness, desire, knowledge, ability, reinforcement
ASEM	American Society for Engineering Management
ASQ	American Society for Quality
BMGI	Breakthrough Management Group Inc.
CAP	change acceleration process
CEO	chief executive officer
CI	continuous improvement
CM	change management
COO	chief operating officer
CPI	continuous process improvement
DISC	a personality profile model
DMADV	define, measure, analyze, design, verify
DMAIC	define, measure, analyze, improve, control
EMO/PERF	emotional response/performance
FMEA	failure modes and effects analysis
HR	human resources
IT	information technology
LSS	Lean Six Sigma
MBTI	Myers–Briggs Type Indicator
OCM	organizational change management
OD	organizational development
PDCA	plan, do, check, act (or adjust)
PM	project management
PMO	project (or program) management office
QWERTY	name of a common computer keyboard design
ROIC	return on invested capital
ROI	return on investment
SQL	Structured Query Language
TPM	total productive maintenance
VP	vice president
TV	television
WIIFM	what's in it for me?

Foreword

Much has been written on change. The sentiments expressed in articles and books that describe the importance and challenges of successful change management are certainly accurate. The statistics often quoted are not. After all, can it really be true that 70% of change efforts fail (a number blindly repeated time and time again)? Probably not. But then again, what exactly does it mean to succeed? Most change DOES eventually happen. The real question is how painful the path turns out to be and how much time, effort, and money is lost along the way. The more time it takes and the more effort required, the less efficient the change effort. Of course, some change efforts DO actually fail, resulting in no net change. In the end, the measurement of change lies along a continuum, and effective change management is about improving the odds of success while reducing the resources required to succeed.

Randy K. Kesterson offers us his career's worth of wisdom. In fact, through research and interviews with others, he offers us many careers' worth. By bringing together the collective wisdom, research, and learning of so many others, Randy creates a truly collaborative compendium with special emphasis on the challenges of driving day-to-day process changes such as those often driven by the continuous improvement leaders we call Six Sigma Black Belts.

Strategic change often comes from high above. While people might drag their feet, mandates from high above generally move forward. In these cases, change management is intended to grease the skids, ensuring change happens as fast as possible.

On the other hand, Six Sigma Black Belts, the principal actors in this book, tend to be operating at peer-to-peer levels, meaning they don't have the authority to mandate change and they don't have the access to the resources necessary to ram change through. Black Belts also tend to be working to improve the daily operating processes of the business, which is much like trying to change the belt on an engine while the engine is running. In other words, for Six Sigma Black Belts, change management efforts are often far more challenging than for their superiors in the organization. Yet collectively, their efforts are no less important. In fact, strategic change driven from on high above is often implemented through the efforts of many Six Sigma Black Belts.

I've known Randy for a decade, and since the day I met him, his passion for understanding and managing change has always been apparent. When we first met, he was at LaMarsh & Associates, a time in his life that he refers to affectionately throughout this book. Ten years on, LaMarsh Global remains one of the top change management firms in the world.

I've seen Randy's passion applied in the workplace when he was a client at General Dynamics. Without his solid background in change management, I don't believe his team or mine would have met with the success they experienced together.

The readers of this book will no doubt be able to relate to many of the experiences and stories Randy shares. And they will likely ask themselves, "Why wasn't I taught this sooner?" The answer is that, while change management is very often referred to as a "soft skill," in practice it is quite hard. Resistance to change is complex and dynamic. Just as a block of wood sliding across a surface meets with resistance from multiple forces, so, too, does effort to change a process. As the block of wood moves faster, resistance increases. The friction presented to the block by air increases exponentially as something moves through it, while the friction between the block and the surface also increases according to the interfaces coefficient of friction. Process change is no different. There is organizational resistance brought on by corporate bureaucracy, there is process friction by virtue of the fact that most processes do not operate independently of all other processes, and then there is the resistance of people who may not understand the need for change or who are just more comfortable with the status quo. Breaking down the change and dealing with the human element is the focus of this book.

David Silverstein, CEO
The Lean Methods Group

Preface

I've had this book in my head for a while. It's been trying to get out, and I finally decided in early 2016 to sit down and start to write it.

The book straddles a possibly untouched niche—the intersection of change management (CM) and Lean Six Sigma. The book is written for the Black Belt, or the organizational change management practitioner, working inside an organization with low Lean maturity, where significant resistance to change is the norm. You see, I believe there is a continuum of organizations with "Fat" (i.e., non-Lean) organizations on the left and "Leaner" organizations on the right.

You may know of organizations with more Lean maturity, where there is no longer as much need for overt change management and Lean has become a part of the culture rather than a set of tools practiced by people in structured Black Belt roles.

A recent list of the top 10 Lean manufacturing companies in the world[1] included the following:

1. Toyota
2. Ford
3. John Deere
4. Parker Hannifin
5. Textron
6. Illinois Tool Works
7. Intel
8. Caterpillar
9. Kimberly-Clark
10. Nike

I'd also nominate Danaher and Milliken for inclusion in the "far right" club.

For the less-Lean organization, building change management tools into Lean Six Sigma projects helps to mitigate resistance to change, accelerating the benefits and saving the organization time, money, and employee angst—in short, eliminating waste. Change management can also help with the Lean Six Sigma initiative deployment or, more likely in today's world, redeployment.

xvi • Preface

Why is it that more organizations are not farther right on the Lean continuum? I have a theory about this. Referring to the elements of Lean on the right side of Figure P.1, where within a typical organization does strategy formulation reside? Typically, at the top, you say, with the executive team or involving a team of experts in the business development function? And who does the lion's share of work on process improvement? Maybe Operations? And who fiddles the most with org structure? Executives with the help of Human Resources/Organizational Development (HR/OD). And who typically strives to improve workplace culture and address nagging culture gaps? Maybe HR/OD again? How about metrics? Who typically owns most of them and tracks them? Maybe Operations and Finance?

The picture I'm painting is that the work to move toward a Lean(er) organization requires the close coordination and alignment of multiple teams of people working across functional areas, up and down the organization. It's not easy. My observation is that organizations that have been most successful (at least large organizations in some parts of the world) have relied upon Hoshin Kanri to create the needed alignment from top to bottom and across the organization.

Some people who have read early drafts of this book come from organizations on the far right side of the Fat–Lean continuum. A few of them

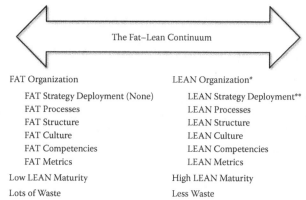

FAT Organization	LEAN Organization*
FAT Strategy Deployment (None)	LEAN Strategy Deployment**
FAT Processes	LEAN Processes
FAT Structure	LEAN Structure
FAT Culture	LEAN Culture
FAT Competencies	LEAN Competencies
FAT Metrics	LEAN Metrics
Low LEAN Maturity	High LEAN Maturity
Lots of Waste	Less Waste

* If you're not familiar with an organization on the right side of the Fat–Lean Continuum, you can get a feel for it by reviewing the Shingo Principles at http://www.shingo.org.
** I'm not suggesting that a FAT organization never *formulates* strategy; I'm saying that most seem to lack an effective strategy deployment and execution process. A strategy deployment approach that aligns very well with Lean and the Shingo Principles is called Hoshin Kanri. You can scan for it on the Internet or, if you want to know more, see *The Basics of Hoshin Kanri*, of Hoshin Kanri (Boca Raton, FL: CRC Press, 2015) a book at the primer level.

FIGURE P.1
The Fat–Lean Continuum. (Created by the author.)

seem to look down their Lean noses at people who don't reside near them in the land of Lean-ness. They tell me that Six Sigma is not really a part of Lean, i.e., you shouldn't require people to work on projects in Black Belt roles, and they say that change management should not be needed. With the greatest of respect, I tell them a story in an attempt to help them to see other (less Lean) perspectives.

An Olympic swimmer was watching a very young boy take his first swimming lesson at the community swimming pool. The boy had finally learned to put his face in the water and was now using a float board while his instructor held him afloat. The Olympian scoffed, "you should not need a float board to swim!" From his position at the far right end of the swimming continuum he had forgotten that he once, long ago, had needed this kind of help as well.

I've spent two-thirds of my career in executive roles in industry and one-third of my time in strategic change management consulting, working deep inside the intersection of Lean Six Sigma and change management. I've helped Master Black Belts within several Fortune 500 companies weave structured change management approaches and tools into the DMAIC (define, measure, analyze, improve, control) and DMADV (define, measure, analyze, design, verify) steps. The results have been significant.

DO NOT READ THIS BOOK!

Do not read this book if you are fortunate enough to work within an organization that lies on the right side of the Fat–Lean Continuum. On the other hand, if you are like most of us, living in organizations on the left side, struggling to move toward the right side of the continuum, facing resistance to change at every turn, please join me for an exploration of the intersection of two fascinating methodologies.

Randy K. Kesterson
Davidson, North Carolina

REFERENCE

1. Top 10 Lean Manufacturing companies in the world, ManufacturingGlobal.com, June 12, 2014.

Working at the Intersection

There are many people in this world who possess more expertise than I do in the field of OCM, and there are also a considerable number of people with far more expertise in Lean Six Sigma.

But, I suspect that I am a member of a fairly small group of people who possess experience and some expertise in both—people with experience working at the intersection of change management and Lean Six Sigma.

Introduction

I'm guessing that you're reading this introduction because you're looking for a simple explanation of Organizational Change Management (OCM). I understand. I sometimes hear people involved in a stalled project tell others, "We need to use change management on this project." But, I wonder, "do they really know what Change Management is?"

This book could be titled "Change Management 101." It is intended to be used as follows:

1. *A refresher* for existing Black Belts, Green Belts, and Master Black Belts to improve their understanding of change management;
2. *A primer* in change management for new belts going through Lean Six Sigma training;
3. *A source of training material* for those leading Lean Six Sigma training classes; and
4. A *source for change management practitioners* coaching existing belts through their projects.

A DEFINITION

"*Change management* is about people. It is the way we help people navigate change successfully. Tools, techniques, and processes exist to help people make the transition to adopt change."

—**Lisa A. Riegel, PhD**
President/CEO
Educational Partnerships Institute, LLC

CONFUSION ABOUT THE TERM

Management of the people side of change is sometimes confused with the term, change management (CM)., used in the parlance of the Information

Technology (IT) professional. In the IT world, the people side of change is critically important, but "change management" is also used to describe an IT service management discipline. The objective of change management in the IT context is to ensure that standardized methods and procedures are used for *changes to control IT infrastructure*—NOT to help people.

I first learned of the term OCM when I was Executive Vice President and COO for an aerospace and defense company. In the many years since, I have learned that change management applies to all forms of change—from breakthrough, cataclysmic change, to incremental continuous improvement projects. In this book, I'll talk about how Organizational change management (OCM) applies to Lean Six Sigma, from the unique perspective of the Black Belt.

TERMINOLOGY

I'll use the terms "organizational change management", "change management", "CM", and "OCM" synonymously throughout this book.

I'll also use the terms, "Lean Six Sigma" and the abbreviation "LSS" interchangeably.

The first part of the book explains the BIG problem—resistance—and why you should care about it. Part II explains what you can do about it. Part III contains Q&A sessions I conducted with experts from the fields of OCM and Lean Six Sigma; many of the interviewees are experts in both; mini-biographies of all the participants are also included. Section IV includes some nuggets of information that should help you as you move on to more advanced change management topics.

Paraphrasing Leonardo da Vinci and Albert Einstein, I hope this book is simple enough, but not too simple.

> Simplicity is the ultimate sophistication.
> —**Leonardo DaVinci**

> Everything should be made as simple as possible but not simpler.
> —**Albert Einstein**

Thank You

To Jeanenne LaMarsh and Jeff Hiatt for sharing a small portion of your change management expertise with me, helping me to understand a few important things about this fascinating subject.

To all of you who spent time with me, sharing your thoughts and answering my questions for incorporation into the "Interviews with Experts" section of this book.

To A. Blanton Godfrey Ph.D. for sharing a bit of your knowledge of Lean Six Sigma with me.

To Ellen Domb, Pierce Howard, Ph.D., Zack Johnson, Alireza Kar, Siobhan Pandya, Lisa Riegel, Ph.d., and Michele Quinn for your valuable contributions to this book.

To Susan and my siblings (Lori Lynn and Kris Anthony) for your support and (mostly) positive, constructive editing.

To Chase for your help with the artwork.

To those who assisted me with the editing of the final draft, you will find your name listed in Appendix F, Acknowledgments.

Finally, to all of you who throughout my career have given me examples of how to (and how not to) effectively manage change.

About the Author

Randy K. Kesterson has held executive-level positions at Cobham, Doosan Bobcat, General Dynamics, and Curtiss-Wright, with prior successful experience at Harsco Corporation, John Deere, and at privately held Young & Franklin/Tactair Fluid Controls.

He also worked as a management consultant to organizations such as Bank of America, Caterpillar, Motorola, Bank of Montreal, Ford Motor Company, Milliken & Company, RJ Reynolds, and the Federal Aviation Administration (FAA).

Randy recently served as the Chair of the Advisory Board for the Center for Global Supply Chain and Process Management at the University of South Carolina's Moore School of Business. He earned his Six Sigma Black Belt at North Carolina State University/IES.

He earned his Bachelor of Science degree in Engineering Operations from Iowa State University and attended Syracuse University, where he earned his MBA with a concentration in Operations Management.

Randy and his family have residences in North Carolina and in the Washington, DC, area.

Section I

1

Why Should You Care?

You can have Change Management without Continuous Improvement; however, you cannot have Continuous Improvement without Change Management. It is a requirement at every stage of a Lean Six Sigma project, and without the acknowledgment of its importance and the deployment throughout the entire project, you will not be successful.

—**Siobhan Pandya**
Director of Continuous Improvement and Lean at Mary Kay Inc. Formerly with Shell Oil Company in the UK and US. Lean Six Sigma Black Belt

You will learn, or maybe you've already learned, that Lean Six Sigma is about changing and improving processes to eliminate waste and/or to reduce process variation. When you change a process, you almost always require people to change what they do or how they do it. Now, there is the rub: This stuff involves people! Oftentimes, people don't like change, even when the change is good for them.

Why should you care about organizational change management (OCM)? Because you will probably fail without it.

Most of the unsuccessful Black Belts I've known have *not* failed because of their inability to use the Lean Six Sigma tools. They've failed because of their inability to deal with the "people stuff," including the inability to deal with the resistance they encountered on their projects.

As a certified Black Belt, or a Black Belt in training, why should you care about OCM?

Years ago, I asked this very question. I simply didn't understand. Now that I have some gray hair on my head, I finally understand that being a good Black Belt requires the following skills along with a set of core competencies. I call these skills and competencies the critical success factors for a Black Belt.

As shown in Figure 1.1, the required skill sets form a triangle, with the core competencies shown to be residing at the very heart of the critical success factors.

Critical success factors = three (3) sets of skills + one (1) set of competencies

The Black Belt must have the ability to use:

1. Lean Six Sigma approaches and tools
2. Project management approaches and tools
3. Change management approaches and tools
4. … and the Black Belt must also possess the required competencies to do the job.

In my experience, the best, most successful Black Belts are masters of the three skill sets shown in Figure 1.1, and they also possess the required Black Belt competencies. You can think of the four elements shown in the diagram below as the critical success factors to be a good Black Belt.

I've intentionally shown the competencies to be in a gray zone, as shown in the triangle in Figure 1.1, because there is disagreement about the makeup of this list. As consultants often say, "It depends." It depends on the environment within which the Black Belt is working—both the project environment and the overall "change receptiveness" of the organization. More on change receptiveness later.

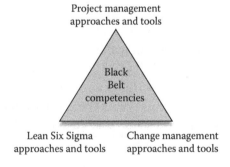

FIGURE 1.1
The critical success factors for a Black Belt. (Created by the author.)

A representation to help explain Figure 1.1 might be:

Probability of Black Belt success = PM * LSS * OCM * Core Comp,

where
PM = a project management skills score
LSS = a Lean Six Sigma skills score
OCM = a organizational change management skills score
Core Comp = A Black Belt competency assessment score

A zero score in any of the aforementioned assessment areas would obviously give the overall probability of success a score equal to zero. In other words, if you don't have some ability in all three skill areas *and* possess the competencies (some of which are congenital, and some of which are learned and developed), you might become a Black Belt, but probably not a good Black Belt.

I believe some of these competencies to be inherent (i.e., if you weren't born with them, you're not going to get them in a training class), while some of the competencies may be shaped by one's environment.

Here is a list of Black Belt competency areas:

- Communication skills
- Analytical, technical, project management skills
- Empowerment
- Passion and enthusiasm
- Leadership, training, and coaching
- Change agency
- Influence, "making things happen"
- Confidence

For those of you who just can't move on without learning more about these competencies, stop now and go to Appendix A for a peek. There you will find more on my personal favorite list of Black Belt competencies. Be advised that the list was not handed down from on high on a stone tablet, so please be aware that others may disagree with the competencies I've included on this list.

Bottom line: if you want to be a high-performing and successful Black Belt, you need to be good at change management. You may not use extensive change management on every project, but if you work enough projects, you will find yourself in need of some knowledge of change management sooner rather than later. At that point, you can choose to handle it yourself with the skills you've accumulated, or you will hopefully be able to recognize the need to call for help from a change management expert (before it's too late).

Sidenote: My friend, Alireza (Ali) Kar of Sydney, Australia, politely reminded me of my sloppy use of the word "probability" in the earlier section. I've left it "as is," because I am trying to make a point without being absolutely pure in the mathematical, statistical sense. I hope those of you offended by this can find it in your analytical hearts to forgive me.

Here are Ali's words:

May I ask, don't disturb the science by creating wrong science. My friend, what does, "Probability of Black Belt Success = PM * LSS * OCM * Core Comp" mean? You failed to understand the meaning of probability. Please read *Fooled by Randomness* by Nassim Taleb to understand the difference between "probability" and "expected value". Thanks, Ali!

CHANGE RECEPTIVENESS OF THE ORGANIZATION

I told you earlier that Black Belt competencies depend on the environment within which the Black Belt is working—both the project environment and the overall "change receptiveness" of the organization. Here's a formulaic way to think about it.

You've heard of $Y = f(X)$, where Y is a function of X?

For example, a person's weight (Y) might be dependent on several factors, the X's.

A person's weight (Y) = calorie intake (X1) * daily exercise (X2) * genetic propensity to be overweight (X3) * number of fast food meals per year (X4), etc.

Several different X's might impact a person's weight, and some of the X's are more correlated with Y than the others.

I propose that the change receptiveness of an organization—let's call it Y—is also a function of several X's, and the outcome of Y is dependent on the X's.

Based on my experience, I'd say if a business unit scores high on the first two Shingo guiding principles—respect and humility it probably also has a more change-receptive environment. Low scores on these two critical "X's," probably not.[1]

Shigeo Shingo (新郷重夫, 1909–1990), was born in Saga City, Japan. He was (and is) considered one of the world's leading experts on what is now known as operational excellence. In 1988, Utah State University conferred an honorary doctorate on Mr. Shingo, a Japanese industrial engineer and author credited for his contribution to many of the elements, theories, and tools associated with the Toyota Production System. That same year, Utah State University established what was then called the Shingo Prize for Excellence in Manufacturing in his honor.

The Shingo Prize for Operational Excellence is now an annual award given to organizations worldwide by the Shingo Institute, part of the Jon M. Huntsman School of Business at Utah State University in Logan, Utah. Criteria defined as a set of Shingo guiding principles are used to select the award winners.

More information can be found at www.shingo.org.

So, the Change Receptiveness of an Organization (Y) = Level of *Mutual Respect* exhibited within the organization (X1) * Level of *Humility* demonstrated by the leaders in the organization (X2) would be a good start if we were looking for the critical X's that drive Y. You will note that I added the word "mutual" to the sentence. I believe that respect must go both ways—from the leaders down to the bottom of the organization and also from the bottom to the top. It's not a one-way street. In my experience, this starts at the top. If leaders show that they respect their people, then they will eventually earn the respect in return.

Respect and humility are cornerstones of a good, solid organizational culture. I think it goes without saying, but I'd throw in the three C's (communication, cooperation, and collaboration) as other critical X's to help make an organization more receptive to change.

Some may be surprised that humility is included as a success factor, since pride and maybe even arrogance are perceived as success factors in a lot of organizations, including some with whom I've worked. I suggest that people at all levels need to be humble enough to recognize that all of our systems and processes can be improved, no matter where they came from.

A CATALYST FOR SUCCESS

There are enablers that help immensely with regard to Lean Six Sigma, OCM, Hoshin Kanri, and other disciplines related to improvement and change. These enablers help foster more of the three C's (communication, cooperation, and collaboration) and form a catalyst, a "spark," that will help ignite the continuous improvement process.

RESPECTFUL AND RESPECTED LEADERSHIP

Using the ten Shingo principles as the starting point, I think of the ideal leader as being both respectful and respected.

The leadership team is respectful of the employees/associates and the employees/associates truly respect their leaders. I believe that leading with humility aids tremendously in gaining employee/associate respect. Command and control leaders and impactful managers will kill a Lean/continuous improvement deployment. Look for the leaders with the "What do you think?" versus the "Thou shalt" approach. Executives need to move the dictatorial leaders and managers out of the way.

\kə-ˌla-bə-ˈrā-shən\

It looks and sounds a bit like the name of a Russian ship, doesn't it? Well, it's not. It's the pronunciation of a very important word for successful change management—collaboration.

THE COLLABORATION TRADE-OFF

A discussion with my friend David Silverstein always results in two things: (1) he says something that would shock even the utmost contrarian; and (2) I learn something. Typically, several somethings.

One day, as I was writing this book, David and I were discussing change management and the topic of collaboration came up. David mentioned a book that he had recently read: *Team of Teams: New Rules of Engagement for a Complex World,* by *General Stanley McChrystal, Tantum Collins, David Silverman,* and *Chris Fussell.*

David explained that he had learned from the book that collaboration is something that involves a huge trade-off. Collaboration improves *effectiveness,* but at the sacrifice of some *efficiency.* In my experience, the collaboration trade-off typically pays huge dividends. In a change project, the gains in effectiveness that come from a collaborative approach are worth the loss of some efficiency, i.e., the project might involve more meetings and discussions, but the enhanced project outcomes will be the reward for your extra efforts.

KEY TAKEAWAYS

- Without OCM, you will probably *not* be successful in your Black Belt career.
- OCM is one of the required skill sets needed by a Black Belt for success.
- In addition to OCM skills, a successful Black Belt also needs
    - Project Management skills
    - Lean Six Sigma skills
    - Certain competencies (some inherent, some acquired)
- Always try to be respectful and act with humility! The Shingo principles help set the tone for any Black Belt, brand new or deeply experienced. See www.shingo.org for more.
- And finally, communicate, cooperate, and collaborate across and up and down through your organization!
- As Ellen Domb recently reminded me, this book is about WASTE (a Lean term)! It is wasteful to have a team do a great job of improving a process, then have the new process not be adopted by the organization, due to lack of understanding of the human side of changing to the new process.

REFERENCE

1. "Shingo Guiding Principles." *The Shingo Model – Shingo Institute.* N.p., n.d. http://www.shingoprize.org.

2

Change Is Good!

It is *really* hard to stay the same size in business. There are market forces always at play. As a result, a business tends to grow or shrink over time. Most shrink and die, or are consumed by larger, healthier organizations. It is not easy to grow a business. It takes skill, good planning, good execution, and a bit of luck does not hurt. If you do not effectively manage a business, it will tend to lose revenue and decline over time.

Processes are similar in this regard. If you leave a process alone, it will tend to decline over time. You have probably heard that the natural state of a process is decay. Decay in this context means to "decline from a state of normality, excellence, or prosperity." Think of a satellite in orbit. Over time, with no outside influence, orbits deteriorate, and the satellite eventually plunges to a fiery death.

To avoid decay, we need to pay attention to the processes that matter. Within the operations part of an organization, that often means seeking to improve things like safety, quality, delivery, and speed. You should have a bias toward action. I have found that it is best to push hard for change, even when others around you do not see the need for it. It is best to make some change, then course correct, whether it is an incremental change or a breakthrough, seismic change. It is almost always better to be changing and evolving versus standing still.

I have always found it helpful to think of myself and the organization as being in a race. I have imagined that the leading competitors in our business niche are slightly ahead of us and that our competitors are right behind us, catching up fast. That mental image has always spurred me on, helping me to avoid any chance of complacency.

So, let us begin this book with the premise that at least *some* change is good, as has been explained by a wide variety of very knowledgeable people from across the reaches of time:

He who rejects change is the architect of decay.

It is not necessary to change. Survival is not mandatory.

—W. Edwards Deming

I cannot say whether things will get better if we change; what I can say is they must change if they are to get better.

—Georg C. Lichtenberg

One must change one's tactics every ten years if one wishes to maintain one's superiority.

—Napoleon Bonaparte

Life belongs to the living, and he who lives must be prepared for changes.

—Johann Wolfgang von Goethe

Change is inevitable. Change for the better is a full-time job.

—Adlai E. Stevenson

Change before you have to.

—Jack Welch

To improve is to change; to be perfect is to change often.

—Winston Churchill

What good is an idea if it remains an idea? Try. Experiment. Iterate. Fail. Try again. Change the world.

—Simon Sinek

Change in all things is sweet.

—Aristotle

The path of least resistance is the path of the loser.

—H. G. Wells

The world hates change, yet it is the only thing that has brought progress.

—**Charles Kettering**

Screw it. Let's do it.

—**Richard Branson**

If "Plan A" doesn't work, the alphabet has 25 more letters!

—**Claire Cook**

It is not the strongest of the species that survives, nor the most intelligent, but the one most responsive to change.

—**Charles Darwin**

When you've finished changing, you're finished.

—**Benjamin Franklin**

KEY TAKEAWAYS

- If you leave a process alone, it will tend to decline over time.
- To be successful, organizations need to work diligently to improve their key processes to prevent this decline.
- As Lisa Riegel, Ph.D. and President/CEO of the Educational Partnerships Institute, recently pointed out to me, "… change is a natural part of growing, and we should embrace advances. We would still be drinking mercury if we didn't apply change to our medical discoveries."

3

Change Is Hard!

Have you ever been on a diet and failed to lose weight? Have you lost weight but failed to keep the weight off? Have you ever tried to stop smoking, but your best efforts (pardon the pun) went up in smoke?

If you answered "yes" to any of these questions, you're not alone. A quick Internet search of how many diet and smoking cessation programs fail most often brings up the statistic of 95% failure rate. Why is this? It's because change is hard! Changing long-established processes (within an organization or within your personal life) is not easy. Changing long-held habits, at home or in the workplace, is just plain hard.

Let me demonstrate.

Do you have a QWERTY keypad on your computer? No, QWERTY is not a slang term used by one of your friends in Great Britain. QWERTY describes the standard layout on English-language typewriters and keyboards, having Q, W, E, R, T, and Y as the first keys from the left on the top row of letters (see Figure 3.1).

When I showed a computer keyboard to my son a few years ago, he asked me, "Why aren't the letters in alphabetical order?" That was a good question, so I googled it.

As a result of that Google search, I learned that Christopher Sholes, the man who is believed to have invented the keyboard, created the QWERTY design *to slow typists down*. You see, the faster someone typed, the more often the typewriter jammed (see Figure 3.2).

According to urban legend, Mr. Sholes put common letters in hard-to-reach spots.

I read recently that this popular theory has been debunked. According to a story in the Smithsonian, the QWERTY keyboard was actually created based on the advice of telegraph operators.[1] The first keyboards

FIGURE 3.1
A QWERTY keyboard. (Stock image purchased from DepositPhotos.com.)

FIGURE 3.2
Photo of jammed typewriter keys. (Stock image purchased from DepositPhotos.com.)

were being used by telegraph operators to translate Morse code, and the keyboards were built for that.

Regardless of how the QWERTY came to be, the time has come, some say, to revise the keyboard for an increasingly mobile world.

But ask yourself, would *you* be willing to adopt a new keyboard? What if you were told that the new keyboard was more ergonomic in design, making you less prone to repetitive motion injuries, e.g., carpal tunnel syndrome, and it might also allow you to type faster? Would you change?

Here is your chance to demonstrate your personal propensity to change! Behold, the Dvorak Simplified Keyboard!

The Dvorak Simplified Keyboard is a keyboard layout patented in 1936 by Dr. August Dvorak and his brother-in-law, Dr. William Dealey. Several modifications have since been designed by the team led by Dvorak. These variations have been collectively or individually called the Simplified Keyboard or American Simplified Keyboard, but they all have come to be commonly known as the Dvorak keyboard or the Dvorak layout (see Figure 3.3).

Dvorak proponents claim the layout requires less finger motion and reduces errors compared to the standard layout, the QWERTY keyboard. It is claimed that the reduction in the finger distance traveled permits faster rates of typing while also reducing repetitive strain injuries, though that claim has been questioned, and that criticism, in turn, has also been challenged.

Here's another opportunity for you to assess your openness to change!

Have you ever heard of the BeeRaider Keyboard? It was created by Ray McEnaney because he wasn't satisfied with the QWERTY or Dvorak. Figure 3.4 shows BeeRaider's optimized radial keyboard. The optimized layout of the alphabet characters on this keyboard is based on the frequency of occurrence of the characters in the English language.

Another keyboard design (see Figure 3.5) is called the KALQ, and it has been designed around a very specific, very modern behavior—typing with the thumbs. Some thumb-typers claim that they can enter text significantly faster using a KALQ design than when using a QWERTY design.

DVORAK keyboard

FIGURE 3.3
The Dvorak keyboard. (Stock image purchased from Alamy.com.)

18 • *The Intersection of Change Management and Lean Six Sigma*

FIGURE 3.4
The BeeRaider (optimized) keyboard. (Image of BeeRaider keyboard provided by Ray McEnaney.)

FIGURE 3.5
The KALQ keyboard. (From Williams, Christopher. Typist-small.png. Digital image. Business Insider. N.p., n.d., Web.)

What's next? Maybe a spherical keyboard? (Figure 3.6)

So, while it might be a great idea, you can see how hard it is to make a change.

According to Jimmy Stamp in his 2013 article about the QWERTY keyboard, "when a design depends on a previous innovation too entrenched in the cultural zeitgeist to change, it's known as a path dependency."

As of this writing, I'm not seeing a lot of thumb-typer or spherical keyboards on computers. Maybe there is something to this path dependency theory?

FIGURE 3.6
The keyboard ball. (Stock image purchased from DepositPhotos.com.)

So, can we move forward with an agreement that at least *some* change is good, and it's also very hard to accomplish change at least *some* of the time? If you're still skeptical, I refer you to comments by a wide variety of very knowledgeable people from across the reaches of time.

> Everyone thinks of changing the world, but no one thinks of changing himself.
>
> **—Leo Tolstoy**

> Change has its enemies.
>
> **—Robert Kennedy**

> Change is hard because people overestimate the value of what they have and underestimate the value of what they may gain by giving that up.
>
> **—James Belasco and Ralph Stayer, *Flight of the Buffalo* (1994)**

> If you want to make enemies, try to change something.
>
> **—Woodrow Wilson**

> There is nothing more difficult to take in hand, more perilous to conduct, or more uncertain in its success, than to take the lead in the introduction of a new order of things.
>
> **—Niccolo Machiavelli, *The Prince* (1532)**

People don't resist change. They resist being changed.

—**Peter Senge**

Change is hard at first, messy in the middle, and gorgeous at the end.

—**Robin Sharma**

It is easier to resist at the beginning than at the end.

—**Leonardo da Vinci**

Change is such hard work.

—**Billy Crystal**

KEY TAKEAWAYS

- Changing long-held habits, at home or in the workplace, is just plain hard.
- Some things become too entrenched in the culture to change. The QWERTY keyboard is a good example of this.
- Changing the U.S. customary measurement system (based on the British imperial system) to the metric system is another good example of a change that is hard. Most countries around the world use the metric system, which uses measuring units such as meters and grams and adds prefixes like kilo-, centi-, and milli- to identify orders of magnitude. The United States has attempted to adopt the metric system, but still widely uses the older, imperial-based system, where things are measured in feet, inches, and pounds.

REFERENCE

1. Stamp, J. Fact or Fiction? The Legend of the QWERTY Keyboard. Smithsonian.com. N.p., 3 May 2013. http://www.smithsonianmag.com/arts-culture/fact-of-fiction-the-legend-of-the-qwerty-keyboard-49863249/

4

Resistance Is Real

Before we get started with this chapter ... *please cross your arms in front of you* (Figure 4.1).

Now, uncross your arms and cross them the other way ... with the *other* arm on top this time.

How does that *feel*? A bit uncomfortable?

Now, uncross your arms and cross them one more time. Did you go back to the original "normal, more comfortable" way? Why?

Change is hard. Even simple changes like that in the arm-crossing example can be difficult. Most of us like the way we do things. That's why we do them a certain way.

We all have resistance to change. At least to some change.

The next time you visit your parents' home, please take a look inside the kitchen cabinets, pantry, and bathroom. Check out the brands of toothpaste, shaving cream, laundry detergent, soap, etc. they're using. Do you see any overlap with the brands in use in your home? Why is this? Resistance to change for many of us can be high. It's so much easier just to stay with what's comfortable. Stay with familiar brands of products. Stay with old habits.

Remember the story of the keyboard in the last chapter? The story tells us that a lot of innovative thinking never gets people to change. Most of us are still using the QWERTY keyboard. Why is this? The answer is, *resistance!*

THE CHANGE CURVE

The following chart is an adaptation from the work done by Kübler-Ross, Daryl Conner, Don Kelley, and others over the years.

FIGURE 4.1
The crossed-arms experiment. (Created by author's son.)

What I am calling the change curve has also been identified, in various shapes and forms, as the "Kübler-Ross Model," "Emotional Cycle of Change," "Grief Cycle," "Five Stages of Grief," "Emotional Response to Change," and "Five Stages of a Change."

The curve shown in Figure 4.2 represents the response that has been found to typically result from what is perceived as a *negative change*.

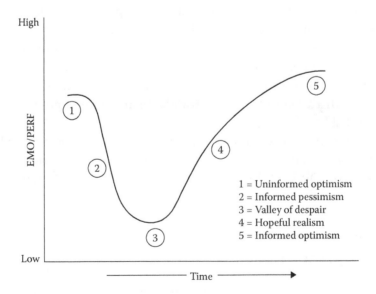

FIGURE 4.2
The change curve. (Created by the author.)

On the *y*-axis in Figure 4.2 is the abbreviation EMO/PERF. This stands for "emotional response/performance." A high-performing team would score high on the *y*-axis, while a low-performing team, as you might expect, would score lower. Likewise, emotional response at the top of the *y*-axis is positive energy, while the bottom of the *y*-axis reflects negative energy. Other words I've seen used to describe the *y*-axis include morale, self-confidence, and perceived effectiveness.

Bottom line, think ☺ at the top of the *y*-axis and ☹ at the bottom.

The *x*-axis indicates the time that passes as the change project progresses.

The five numbered balls in Figure 4.2 indicate stages one moves through when faced with a negative change—or at least one that is perceived negatively by the target of the change. Keep in mind, this change may be considered a *positive* change by leadership, management, and the project team. The target is the person who is impacted directly by the change. The change will require him or her to do something differently in the future, that is in a future state. More on targets later.

As you can see, emotional energy and performance both decline, and then eventually return to normal, (or, ideally, to an even *better* place than before) once the person going through the change has departed from the so-called "valley of despair."

People resist change, or they resist being changed. Either way, resistance is a big deal (see Figure 4.3).

FIGURE 4.3
Resistance. (Created by the author.)

KEY TAKEAWAYS

- How did you do in the arm-crossing example? Did you go back to your original way? Probably so.
- Resistance is real.
- What keyboard do you use? Is it the QWERTY? Probably so.
- Resistance is real.
- Have you, or someone you know, lived through the change curve based on the Kübler-Ross model?
- Resistance is real.

5

Minimizing Resistance

How does one deal with resistance to change? Is there a way to eliminate resistance or to at least mitigate it? You're probably aware of structured *problem-solving* approaches. Lean Six Sigma includes many approaches and tools intended to achieve and enable problem solving. You will probably be pleased to know that there are also structured approaches to eliminate, or at least reduce, resistance to change.

I will tell you a story that literally changed my life. It significantly altered my career path in a good way, and it changed my perspective forever on how to successfully lead change projects, regardless of size and scale.

THE STORY OF THE SHOE THAT CHANGED MY PERSPECTIVE ON CHANGE

Several years ago, I was flying home to the United States from Europe. It was a Friday, at the end of a long week of business travel. In between the bumps and lurches of the plane, I was working to put together a plan to deploy and execute major change initiatives across our company. The first attempt was pretty simplistic:

1. *Formulate strategy*—Develop a differentiating strategy that results in a set of strategic objectives for the business; and
2. *Deploy the strategy*—Carry out the strategy by cascading objectives down into the units and functions and by launching initiatives to attack the biggest problems, improve the key processes, and improve the overall business results.

We were using a few tools at that time:

Tools to formulate strategy—A number of strategy formulation tools, including the Balanced Scorecard, along with some highly paid consultants. I learned about Hoshin Kanri later;

Tools to deploy the strategy—The Balanced Scorecard, Lean Six Sigma, and Project Management.

I remember sketching out our "Current State" on an airline napkin (literally). I will call what we were doing back in those days, "Version 1.0" of an evolving strategy formulation and deployment approach. We had a "tool box" with some tools (see Figure 5.1). Period. End of story.

During this trip to Europe, I had met with the managing director of a company we had recently acquired. Gerhard and I had talked extensively (well, in retrospect, I talked a lot while he listened politely) about making significant changes within his business unit. I remember having the feeling as I was driving back to the airport in Zurich that the trip had been an utter failure. Gerhard was going to keep things the way they were before we had acquired his company. I remember reflecting on how the resistance to change was incredible, especially at some of our newly acquired business units.

But, we couldn't give up! Gerhard's business unit's return on invested capital was nowhere near our targeted level, and it was imperative that we (I) fix this. Somewhere over the North Atlantic, after a couple of glasses of Cabernet, I remember drawing another diagram on the back of another napkin, and this diagram consisted of a few more boxes. I'll call this Version 1.1 (see Figure 5.2).

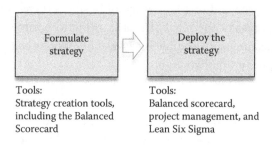

FIGURE 5.1

Version 1.0. Strategy formulation and deployment approach. A toolbox containing a few tools. (Created by the author.)

Minimizing Resistance • 27

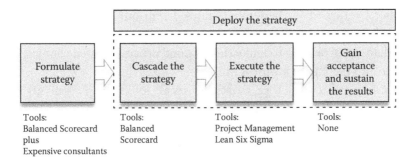

FIGURE 5.2
Version 1.1. Strategy formulation and deployment approach, with more details added on the deployment side. (Created by the author.)

The process I had drawn consisted of four major steps, with "deploy the strategy" now broken into three pieces:

1. *Formulate strategy*—Develop a differentiating strategy that results in a set of strategic objectives for the business;
2. *Cascade the strategy*—Flow the strategic initiatives (that came from the strategic objectives) down into the organization (via numerous scorecards);
3. *Execute the strategy*—Attack the biggest problems, close the biggest gaps, improve key processes, and improve the overall business results; and
4. *Gain acceptance and sustain the results*—Truly engage with the employees [engage with "their hearts and minds," as our human resources (HR) people would say] to eliminate resistance to the changes and improvements and to sustain the hard-fought gains once they have been achieved.

I listed the tools we were using at that time in support of each of the four steps:
  Tool to formulate strategy—Some strategy creation tools, including the Balanced Scorecard (along with some highly paid consultants)
  Tool to cascade the strategy—The Balanced Scorecard
  Tools to execute the strategy—Lean Six Sigma and Project Management
  Tools to gain acceptance and sustain the results—Nothing. Nada. Zip. We had no formal approach nor any tools to help us in this area.

After I returned to the United States, I sent our best Lean Six Sigma Black Belt, a guy I will call "Ph.D. Stan," to Switzerland to see what he could do about helping to make the required process-related changes happen within Gerhard's business unit.

Fast-forward a couple of months. I recall looking up from my desk and seeing Ph.D. Stan standing in my office doorway. Stan had his head down, and his body language indicated that he was thoroughly dejected. "They don't want to do it," he said. He went on to explain that he had used every analytical arrow in his analytical quiver, but he could not convince Gerhard, the managing director, and his team to make the necessary changes to improve their key business processes.

A few days later, I explained my problem to Martin, our VP of HR. After listening intently (as good HR people tend to do), Martin suggested that what we needed was organizational change management. He called it "OCM". I thanked him, but given its roots in the organizational development (OD) world, I assumed this "OCM thing" was grounded in a lot of hand-holding and singing of kumbaya. It probably even included the use of the dreaded F-word, highly promoted by the HR and OD world... *feelings*.

However, I did some research and found that OCM was being employed successfully in industry. In fact, General Electric had been using an approach called the change acceleration process in conjunction with Six Sigma for several years with noteworthy results.

I remember thinking that we needed a similar approach—a methodology, model, toolset that was synergistic with strategy deployment and the Lean Six Sigma tools, something we could use to eliminate the tremendous resistance to change we faced in Europe and in other "foreign thinking" parts of the business, e.g., California.

Just kidding, California... sort of.

My conclusion at the time was that we were using the Balanced Scorecard to develop "balanced" top-level objectives, and we were also using it to some degree to deploy strategy by cascading the scorecards down into the organization. We were also using the Lean Six Sigma tools (in some areas anyway) pretty successfully. But the thing we lacked was a tool set—a methodology, an approach—that could help us deal with resistance to change. We needed something that would work well with the Balanced Scorecard and Lean Six Sigma and that we could deploy across the enterprise.

After doing quite a bit of research about OCM, I concluded that an OCM tool, approach, or methodology was needed to help eliminate resistance to the major changes we were trying to make, to help gain adoption, and to help us sustain the results.

I scheduled a meeting with my boss (Gerry) to talk about the plans for the proposed new strategy formulation and deployment approach. I also invited a consultant who was working with us at the time, who happened to be my Lean Six Sigma Black Belt instructor from years earlier.

I explained the four-step model to my boss: strategy formulation using some strategy creation tools, including the Balanced Scorecard; cascading of the strategy using the Balanced Scorecard; strategy execution using the Lean Six Sigma tools and Project Management; and Resistance Management, using an approach called "organizational change management." With that, Gerry held up his hand to indicate that I should stop talking. He said, "I have a change management approach that has served me very well for over 30 years in business." I can still see my consultant friend lean forward to listen more intently, as he has always been eager to learn about new, successful approaches. My boss then swung his foot up on the conference table, pointed to it, and said "11-E" (his shoe size). "If they don't want to do it, I kick them in the butt," he said. Gerry called it his "11-E approach to change management." I will say that while my boss at the time had indeed perfected the art of 11-E, he was saying this mostly in jest (see Figure 5.3). The meeting ended soon thereafter without a positive resolution (at least from my perspective).

I left that company soon after and was subject to the restrictions of a noncompete agreement that prevented me from working for any company in the industry with which they competed. I knew I couldn't just sit at

FIGURE 5.3
The 11-E approach (11-E was the size of my boss's shoe.) . (Used this photo in my The Basics of Hoshin Kanri book.)

home and do nothing, so I decided to go into the world of management consulting, specializing in an area I wanted to learn more about. You guessed it—OCM.

> **Note to Reader**
>
> *There are times when the 11-E approach is needed … A fire in the hallway doesn't call for consensus building. It's the type of change that calls for immediate action! "The building is on fire, get the hell out of here!"*
>
> *And, there are also times and places for command and control management. For example, can you imagine a military environment without a command and control approach? But, when you're trying to drive strategy down into an organization, for example, when using a Catch ball process,* a command and control approach just doesn't work very well.*
>
> I knew from experience that the 11-E approach to change management was not entirely effective, so I went in search of a new, more enlightened approach. I intended only to remain in this field for a short time, but I enjoyed it so much that I continued on that path for almost 4 years. The wealth of information that I learned in those 4 years about managing resistance and sustaining change proved extremely valuable later in my career.

KEY TAKEAWAYS

- There are structured approaches to eliminate, or at least reduce, resistance to change. Companies such as General Electric have used them for decades, with great success.
- Most organizations have approaches and tools to help them create, deploy, and execute strategy. Strategy execution often involves significant changes within, and sometimes throughout, the organization. But many of these organizations do not have approaches and tools to help them manage the resistance to these changes.
- The 11-E story is about one of these organizations.

---

* See my book, *The Basics of Hoshin Kanri*, of Hoshin Kanri (Boca Raton, FL: CRC Press, 2015), for much more on the Catch ball process and other related subjects.

6

Why?: The Most Important Word in This Book

After living through the 11-E story, I became fascinated with change management, and I studied it in earnest for the next 4 years. I read every book and article I could find on the subject of change management. I spoke with John Kotter, author of *Leading Change*, on the phone. I wrote a couple of articles, of which one was published in the *Wharton Leadership Digest*. I eventually became certified in two organizational change management (OCM) methodologies and met two incredible thought leaders in the field. I studied under Jeanenne LaMarsh and rode horses with Jeff Hiatt as part of his training and certification program in Boulder, Colorado.

Along the way, I learned that successful change management almost always requires the completion of an initial step. It's only one step, but it's a BIG, very important step. If you remember nothing else from this book, please remember this one, simple word.

WHY?

From all my personal experiences and all of my studying and training in OCM, I've learned that people must understand *why* a change is needed as a first step to accepting the change.

People must develop an *Awareness* of the need to change. Once aware of the need, they must also have some personal *Desire* for the change.

To build awareness and to create desire, it often helps to explain the business case for the change, and most people also want to understand the WIIFM, i.e., the What's in it for me?

All of these have to do with the Why.

Until you explain the Why, you should *not* attempt to explain the What or the How or the Where or the Who or the When. Most people won't even listen to you. They're still living back in the current state (to be explained later in this book) asking, "But Why do we need to change? Why do I need to change? What's the business case for change? What's in it for me? Where's the WIIFM?!?!?"

The problem with many project teams is that they've had these discussions amongst themselves for a while (sometimes months), and they've finally convinced themselves about the Why. They have Awareness of the need for change, and they have the Desire to make the change. They understand the business case, and they personally understand the WIIFM. But what do many of those project teams do? They often start the discussions with others who are outside the team by explaining the What (what we are going to change) and the How (how we are going to do it), while the people outside the project are still asking themselves, "Why do I want to do this?"

It typically takes many attempts, using various communication techniques, to convince people of the Why. In my experience, you rarely convince everyone, but once you've convinced enough of the people, you're finally ready to move on to explain the What and the How.

WHAT? What are we going to change?
HOW? How are we going to do it? What's the plan?

THE TARGETS OF THE CHANGE

An important group of people outside the project team are the people who will need to change to make the project successful. These people are the so-called "targets" of the change.[1] The targets of the change also ask, "Why do I want to do this?"

A lesson I've learned is that if you haven't convinced enough people of the Why, and you move forward anyway, you move forward at your project's potential peril. I speak with experience. I've done this, and I've failed. I also learned that it's not all about numbers. You must convince the people who are the influencers within an organization, and oftentimes, these people are *not* in formal positions of power. The influencers are the people to whom others listen and respect. I'm guessing you have observed some influencers in organizations where you have worked. Convince the influencers and you're on your way to satisfying the Why!

PEOPLE ANALYTICS TOOLS

A few years ago, I discovered a neat new tool that I believe could prove invaluable to large organizations in answering the question, "Who are our influencers?" Who are those people, regardless of their formal authority in the organization or lack thereof, who have "the ear" of their coworkers? And sometimes the ears of the coworkers are thousands of miles away.

If an influencer *likes* and supports a change, good things tend to happen for the project team. They tell others, and/or others ask them for their opinion about a change. Organizations love to find these people. For example, if Sally or Bob like this new <fill in the blank>, a lot of people who are sitting on the proverbial fence will too.

The tool is based on an approach used initially to find "the bad guys" in the outside world. If person X is known to be involved in criminal or terrorist activity, then the good guys want to learn with whom person X is connected. With whom is person X communicating? By phone, by e-mail, by text, by other electronic means?

Some large organizations have taken this approach inside, and they are learning that Sally acts as a major hub of communications, i.e., she is widely connected in the organization's communication web, via company-owned communication devices and networks, to thousands of people. What Sally thinks matters to a lot of her coworkers. If we can make Sally *aware* of a major change, and gain her personal *desire* to become involved, good things can happen for the project team (Figure 6.1).

FIGURE 6.1
An influencer within an organization. (From "4 Tips to Becoming an Influencer", Blog article by Marc Guberti, August 22, 2014.)

Sidenote: Mapping influencers is just one of the many ways that analytics are changing how managers attack change. Over the coming years, People Analytics will provide managers with a new toolset that will make process and behavior change easier to track, incentivize, and make a core organizational capability.

Today, tracking behavior change is focused on measuring results, which, as anyone who has worked in an organization before knows, can be way more subjective than it should be. While an important yardstick for measuring change, only measuring results misses a whole lot of the 'why.' It's sort of like treating an aching arm with painkillers as opposed to resetting a bone. If you don't know the driving behaviors that are helping or hindering results, then you will miss what really is or isn't changing.

People analytics allow organizations to measure and optimize the behaviors that drive successful outcomes. The category encompasses a broad swath of behaviors, including measuring who communicates with who, how they communicate, which medium, how communication connects into a broader network, how this impacts things like retention, customer success, sales, and far beyond.

People Analytics are particularly powerful for change because they allow managers to track who is actually changing daily behaviors and who is not. You can see if someone is communicating with the groups they are supposed to be communicating with. You can see if someone is logging into the system they are supposed to be logging into and how that's impacting team behavior. You can see how this impacts onboarding times because every aspect of the employee process is measured.

With tracking also comes incentivizing. You can game behavior change, and make sure people who adopt early are rewarded, both publicly through recognition and with tangible rewards. If seamlessly embedded into work processes, this system could theoretically drive near continuous change that is simply embedded into day-to-day work.

At the end of the day, using these techniques require experimentation, strong values, and good governance. You don't want to look at all data because that's creepy. You don't want the data to by accessed by anyone in the wrong capacity. You also want to make sure the learning from the data is synthesized into action, where it is used either to continue or to kill the change.

People analytics present powerful ways to enhance the change process, and we will see much more in the years to come.

— **Zack Johnson**
Co-founder & Former CEO,
Syndio Social

Syndio Social was acquired by Edge Analytics in 2016. See www.synd.io for more information.

KEY TAKEAWAYS

- WHY is the most important word in this book. Seriously!
- People must understand *why* a change is needed as a first step to accepting the change.
- Next, people must develop an *Awareness* of the need to change.
- Once aware of the need, these people must also have some personal *Desire* for the change.
- Until you explain the Why, you should *not* attempt to explain the What or the How or the Where or the Who or the When. Most people won't even listen to you.

REFERENCE

1. Potts, R., LaMarsh J., Change—The Basics. Master Change, Maximize Success, Chronicle LLC, San Francisco, CA, 2004. 31. Print.

7

Resistance in Your Personal Life?

The following information may be a bit painful to consider, but it's been included to help bring the understanding of resistance home, so to speak.

Have you seen the TV commercials about children starving in other parts of the world (Figure 7.1)? Have you seen the commercials about pets in need of adoption to prevent euthanasia (Figure 7.2)? In both cases, these commercials are making a personal appeal to *change* you, to *change* your behavior.

Have you responded by donating money? If not, why not?

The commercials are intended to make you *Aware* of the problem, but you may not have the *Desire* to become personally involved.

This is a personal example of living in the first two steps of the Prosci ADKAR Model®[1] for organizational change management—Awareness and Desire.

More on this in Chapter 8.

KEY TAKEAWAYS

- Every one of us has resistance in our life: at work, at home, and everywhere.
- If you observe very carefully (and honestly), you will see yourself resisting some changes that you encounter and you will observe others doing so as well. It's a part of life.

FIGURE 7.1
Starving children. (Permission requested from scmun.org to use "starving children" photo.)

FIGURE 7.2
Homeless dog. (Permission requested from TopofOhioPetShelter.org to use "homeless dog" photo.)

REFERENCE

1. Prosci, ADKAR, and Awareness Desire Knowledge Ability Reinforcement are registered trademarks of Prosci, Inc., used with permission.

Section II

8

What Is Organizational Change Management?

A structured process for supporting individuals and groups from the current state to a new future state to deliver organizational results.

—**Michele Quinn**
Certified Prosci Change Management Practitioner, Prosci Master Instructor and Train-the-Trainer professional, Former LSS Deployment Leader, and Master Black Belt

I met Michele Quinn many years ago when she was teaching OCM to a group of Black Belts in training. Her credentials are fairly unique in that she is a certified Master Black Belt, and she's also certified in change management.

Michele's definition of OCM aligns perfectly with Lean Six Sigma in that they are both *structured processes* that, when used properly, *deliver organizational results*.

The OCM approaches I learned, and became certified in, are *not* touchy-feely nor kumbaya- like. They are fact based and structured, and they rely heavily upon data. For many Black Belts in training, this probably sounds great. The problem is that much of the world's population doesn't think this way, so we need to take these differences into account as we are working with teams of people to change what they do and/or how they do it. More on this in Chapter 13, "People Are Different."

> **A**wareness **D**esire **K**nowledge **A**bility **R**einforcement

FIGURE 8.1
Prosci ADKAR Model®. (Image Source: Prosci ADKAR Model ©2016 Prosci, Inc. All rights reserved; Permission to use ADKAR Model info granted by Prosci.)

THE ADKAR MODEL (BY PROSCI, INC.)

The first OCM approach I will explain was designed to deal with individual change. It's a step-by-step method developed by Jeff Hiatt and Tim Creasey at Prosci Research. It's called ADKAR, an acronym that includes the words Awareness and Desire as the first two steps in the five-step process.

Similar to the DMAIC process used in Lean Six Sigma, with ADKAR you move through five distinct steps as you work through a change.

Awareness of the need to change
Desire to participate in the change
Knowledge about how to change (skills and knowledge needed)
Ability to implement new skills & behaviors (ability to perform the skills and act on the knowledge)
Reinforcement to keep the change in place (incentives)

The ADKAR model is shown in Figure 8.1.

Most Lean Six Sigma projects involve more than one person. You're often dealing with groups or teams of people.

MANAGED CHANGE MODEL (BY LAMARSH GLOBAL)

Another approach and methodology for change management is called Managed Change, and it was developed by Jeanenne LaMarsh.

The LaMarsh approach to change management has three primary functions: first, is to identify who, among those who will be impacted by the change, is likely to resist it; second, to assess the sources, types, and degrees of resistance the change may encounter; and third, to design effective strategies to reduce the resistance (Figure 8.2).[1]

What Is Organizational Change Management? • 43

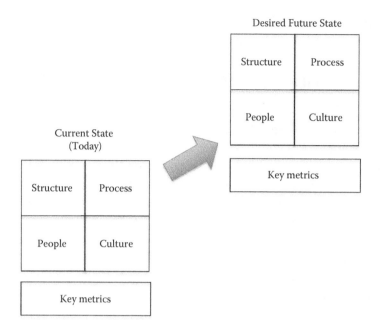

FIGURE 8.2
LaMarsh's Current State/Future State model. (Permission to use LaMarsh Model info granted by LaMarsh Global.)

The Current State can be defined in terms of five elements: structure, process, people, culture, and key metrics.[2]

Structure = geography, organization, systems
Process = the way we do work
People = skills and competencies
Culture = beliefs, behaviors, rules (written and unwritten)
Key metrics = key measures for this part of the business

Resistance to Leaving the Current State

When trying to make more complex change, resistance can come from people not wanting to leave the Current State. These are people who are satisfied with the status quo, who are asking, "why change?"

Resistance to Moving to the Future State

The Desired Future State can be defined using the same approach, in terms of structure, process, people, culture, and key metrics. Some people who are ready to leave the Current State will not be willing to move to the Future State as defined by the project team. They don't want to go there.

Resistance to the Change Plan

Some people will not like the plan that has been constructed to move from Current to Future State. They don't want to follow the plan defined by the project team.

Most Lean Six Sigma projects focus largely on the process and that generally reduces the scope of the resistance faced by the project team.

I'm not going to attempt to make you an expert in ADKAR or the LaMarsh Managed Change Methodology in this primer level book. For that, you'll have to dedicate some time to taking training classes and then employ what you learned in the real world. Appendix E in this book includes information on where to go for more information about OCM approaches and tools.

There are several good change management approaches and methodologies available to you. Prosci, for instance, has developed change models in addition to ADKAR. I am qualified to write about only the two for which I was trained and certified. Chapter 15, "Interviews with Experts," includes comments about other OCM tools.

Remember the change curve you saw in Chapter 4 (Figure 4.2)? That same curve is the solid line in Figure 8.3. The solid line represents the

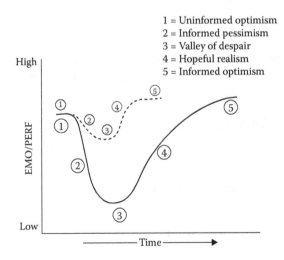

FIGURE 8.3

The change curve: with and without a well-managed change. (Created by the author.)

traditional change curve when there is no application of change management to the change. Figure 8.3 also includes a curve constructed out of a dotted line, which is meant to the show the result of a well-managed change. It's just a representation, but the assumption here is that the use of appropriate, and appropriately timed, OCM approaches and tools will lessen the "dip" in the curve and shorten the amount of time that the targets of the change spend in the "valley of despair," as well as reduce the amount of time it takes to complete the entire project.

So, appropriately applied OCM approaches and tools lead to

- Less time required for the change project
- Less negative energy expended
- Less disruption to the organization
  - Less "dip" in emotional response
  - Less "dip" in performance
- A greater chance of the project team making it to the Desired Future State
- A greater chance of sustaining the gains once there

THE FIVE STAGES

The *Kübler-Ross model*, or the *five stages of grief*, postulates a series of emotions experienced by terminally ill patients prior to death. The model was first introduced by Swiss psychiatrist Elisabeth Kübler-Ross in her 1969 book *On Death and Dying** and was inspired by her work with terminally ill patients.

The five Kübler-Ross stages are as follows:

Stage 1: Denial
Stage 2: Anger
Stage 3: Bargaining
Stage 4: Depression
Stage 5: Acceptance

Don Kelley and Daryl Conner developed their emotional cycle of change model in the mid-1970s, and they outlined it in the *1979 Annual Handbook for Group Facilitators*.

The Kelley/Conner cycle has five stages as described below:

Stage 1: Uninformed optimism
Stage 2: Informed pessimism
Stage 3: Hopeful realism
Stage 4: Informed optimism
Stage 5: Completion

*Note: Kübler-Ross stated later in life that the stages are not a linear and predictable progression, and that she regretted writing them in a way that was misunderstood. Rather, they are a collation of five common experiences for the bereaved that can occur in any order, if at all.

What's shown in Figure 8.4 is an amalgamation of several models that all stem from the *Kübler-Ross model*, or the *five stages of grief*, and/or work by Don Kelley and Daryl Conner, Jeanenne LaMarsh, and possibly others. The takeaway is that most people go through stages when dealing with change and change management, which when done well, *help lessen the depth of the "dip" in the curve and can also reduce the time spent* in what some call the "valley of despair." Jeanenne LaMarsh called this the Delta State, a place that lies between the Current State and the Desired Future State.[3] The results of a well-managed change are also less time spent, less project cost, less disruption, higher performance sooner ... less all-around angst.

The table in Figure 8.4 shows the five stages of change, along with terms sometimes used to describe them. There have been many versions of,

Stage of the Change	I	II	III	IV	V
Descriptions sometimes used to describe each stage	Uniformed Optimism	Informed Pessimism	Valley of Despair	Hopeful Realism	Informed Optimism
Terms used by Jeanene LaManh to describe the progression through a change	Current State	The Delta State			Desired Future State
Some words used to describe things at each stage	Status Quo Unaware Naïve Uninformed Unaware	Shock and Denial Disbelief, Frustration Avoidance, Irritation Confusion, Anger Fear, Uncertainty Bargaining Struggling Pessimistic Aware of the change... ...but no Desire Do Accept it	Depression Overwhelmed The Blahs Lack of Energy Helplessness Fight Regret Despair Giving Up Low	Acceptance Openness Discovery and Exploration Enthusiasum and Hope Moving on Encouraged Desire to Help/Change Seeking Knowledge and Ability	Engagement Acceptance Commitment Sucess Fulfillment Improved Performance High Performing Enpowerment Self Esteem
How to help	Informaton and Communication			Clear Direction	Involvement and Encouragement
		Emotional support			

FIGURE 8.4
The five stages of change. (Created by the author.)

what this author calls, the "change curve" presented over the years, so I've attempted to capture some of the variations in a table form.

As you can see in the table, Jeanenne LaMarsh's Delta State is shown to span stages II, III, and IV. The table also includes some terms used in various "change curves" readily available on the Internet.

The bottom of the table shows some high-level "how to help" guidance as a person moves through the stages of change, starting with providing information and communication, moving on to emotional support, and finally to guidance and direction. Attempts to help must be aligned to where people are in the change.

There is no specific "recipe" for all of this, as to how much of what approach or tool to use and exactly when to apply it. What I've found is that every change project is different, and different people go through the same change in different ways and at very different speeds.

It's a surprise to some that everyone doesn't go through the stages at the same speed. Some move swiftly from the Current State to the Future State, with barely a glance at the Delta State. But others will progress at their own pace—some slower, some very slow. Depending on the change, some will *never* be able to leave the Current State. They might give up. They might quit. Or, they just might stick around and agitate others, either directly or passive-aggressively, to undermine the change. In those situations, you need to get management involved to help deal with the active or passive resistance.

EMOTIONAL SUPPORT

Some who helped me with the editing of early drafts of this book were troubled by the term "emotional support" in a business setting. They'd ask, "What are you recommending, that the Black Belt provide psychiatric counseling?!?!" My response: "These are not couch sessions or ice cream socials. They are merely opportunities for back and forth conversations to improve understanding. If more is needed, seek out your OCM, HR, or OD expert for help."

A Chaordic Change, Discovered in an Unlikely Place

I recently became aware of a significant change project that didn't involve any application of Lean Six Sigma tools, yet the story fits this book perfectly. It's a story about a church in Alexandria, Virginia, on the outskirts of

Washington, DC, which recently embarked on a change that would impact thousands of people. After 60 years of worship in a stately sanctuary, the old building was finally starting to show wear and tear. A decision was made after years of contemplation and praying to renovate the facility. That meant moving two Sunday worship services out of an ornate sanctuary with lofty ceilings, stained glass windows, and wooden pews into a nice, but frankly fairly ordinary, fellowship hall for the 18 months that would be required for demolition and refurbishment. The fellowship hall, replete with linoleum-tiled floor and suspended ceiling blocks, had the look of a place that could have once been used as a gymnasium—not very holy looking indeed. You can imagine the resistance some felt to no longer attending church services in a place that had come to be so special and so familiar.

I was there in attendance the first week the church services were held in the fellowship hall. It was a significant change, and yet it went very, very well. It went so well because of all of the countless hours of detailed planning done by the pastor and his team of devoted parishioners. It was a significant change, but it was handled well because, whether the parishioners knew it or not, they had done a fantastic job of change management. The people were told of the problem. They were made aware of the need to renovate the church. Communications were held every week, using different messengers and various forms of media to explain the change and to build personal desire to participate in the change. People came to understand the business case. They came to understand the WIIFM (the What's in it for me?).

Pastor Don capped off the communication campaign by speaking about the subject of change during that first week in the temporary sanctuary, and his message was about the word Chaordic: "a system or organization that blends characteristics of chaos and order." Sitting there that Sunday morning on a plastic chair, on what could have once been a basketball gymnasium floor, I had the sense that this could have REALLY been a chaotic morning—and yet I was surrounded by order. Pastor Don and his talented team had, through superb planning and seemingly flawless execution, achieved a blend of chaos and order. I can't think of a better way to describe the Delta State when living through a well-managed change. Huge kudos to Pastor Don and his dedicated team!

> I give credit to Chief Pastor Don Davidson for introducing me to the term, Chaordic—a blend of chaos and order. The term was coined by Dee Hock, the founder and former CEO of the VISA credit card association.
>
> **—From Wikipedia**

WARNING!

For those who like engineering stuff and equations, please read on. For those who don't, you may want to go directly to the Safe Zone sign, shown in the next section.

FRICTION EQUATION

Another way to explain the value of OCM involves the equation for determining the resistive force of friction.

$$F_F = \mu N$$

where
  F_F = resistive force of friction
  μ = coefficient of friction for the two surfaces
  N = weight of the object

If a block of aluminum (shown in Figure 8.5) weighing 1 lb is pulled along a dry, clean aluminum plate, it requires approximately 1.2 lbs of force to make the block move. This assumes that (μ), the coefficient of friction between the two surfaces equals 1.2. If you lubricate the surfaces with oil, the force required to make the block move drops to around 0.3 lbs, given that (μ), the coefficient of friction has been reduced to 0.3. In this example, it's four times easier to make the block move after lubrication, i.e., resistance drops significantly after you apply oil (Figure 8.6).

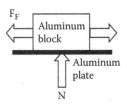

FIGURE 8.5
An aluminum block on an aluminum plate. (Created by the author.)

50 • *The Intersection of Change Management and Lean Six Sigma*

FIGURE 8.6
An oil can. (Created by the author.)

ORGANIZATIONAL CHANGE MANAGEMENT

OCM is like "oil used to reduce the friction between two surfaces." It can help you identify and mitigate resistance to (1) the overall Lean/Continuous Improvement initiative and (2) the improvement projects that will result from the deployment.

Some Lessons Learned on How Can I Help?

Recognize the loss of control that is perceived by some employees
Increase your communication; informed employees are more capable of coping
Don't over promise; you can provide clarity, but you cannot provide certainty
Manage your own demeanor and behaviors; the organization is looking for grounding
Think hard about the concept of "honesty"; be honest, not familial.

Years ago, I attended a conference at Eli Lilly where Brad Claretto, a Master Black Belt, and expert an OCM, presented. My notes from the conference included the aforementioned lessons learned.

The Reinforcement Paradox?

Some have asked me, "Don't the terms Reinforcement and Control fly in the face of change?" or "Doesn't the (R) Reinforcement step in ADKAR and the (C) Control step in DMAIC actually work to *prevent* future change?" or "Doesn't the drive to reinforce and control the new process 'lock things in' forever?"

I tell them that continuous improvement is like a never-ending stairway to excellence. You take one step up, reinforce it to keep the process in control, then when ready, take the next step up and repeat the process.

In Figure 8.7, we start with an important process that has poor results. Let's say the key metric in the y-axis represents first pass yield. The x-axis represents time. We're starting with only 25% first pass yield, and as you can see from the process control chart, the process is not in control. The results of the process vary quite a bit and are not predictable. On the basis of our project selection and prioritization process, we decide to improve this process, so we launch a change project, using the Lean Six Sigma tools.

CHANGE #1

As a result of Change #1, the results improve to 60% first pass yield and the process is now in control! That's a big improvement and we "lock in" the process to make sure we retain those results. In DMAIC vernacular, that is "C", the control phase. We want to keep the process in control. In ADKAR terminology, we want to "R" (reinforce) the new current state. We don't want to drift back to old, bad practices and allow first pass yield to erode. Note that OCM helps to eliminate resistance so we can make the change, and it also helps with *sustaining* the gains once we have them.

But, 60% first pass yield is still not very good. We decide (based on priorities) to launch a second change project to see if we can improve the process further.

CHANGE #2

As a result of our second project, first pass yield is increased to 90%. And, the process remains in control. We might choose to launch another change project later to attempt to improve the process again. It all depends on how improving this process compares in priority to other improvement work we could do with the resources we have available.

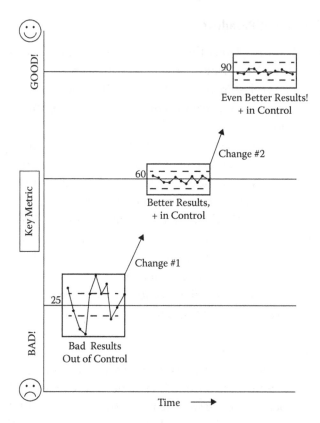

FIGURE 8.7
The never-ending stairway to excellence. (Created by the author.)

So, it starts with identifying a key process in need of improvement. We use Lean Six Sigma tools to improve it, then lock it in. Later, we improve it again and lock in the results. Later, we repeat the process again. And again, and again, and again … That's why it's called continuous improvement.

KEY TAKEAWAYS

- To paraphrase Michele Quinn, OCM is a structured process used to deliver organizational results.
- Prosci's ADKAR approach and LaMarsh's Managed Change methodology are examples of proven, structured OCM processes.

- OCM approaches and tools, when used correctly, will lessen the "dip" in the Change Curve and shorten the amount of time that people spend in the "valley of despair."
- OCM also helps to reduce the amount of time it takes to complete the entire project.
- Five Stages of Change have been defined to explain the steps a person typically takes as they move through the "change curve." This concept was originally described by the Kübler-Ross model, or the five stages of grief.
- Appropriately applied OCM approaches and tools lead to
  - Less time required for the change project
  - Less negative energy expended
  - Less disruption to the organization
    - Less "dip" in emotional response
    - Less "dip" in performance
  - A greater chance of the project team making it to the Future State
  - A greater chance of sustaining the gains once there

REFERENCES

1. Potts, R., LaMarsh J., Change—The Basics. Master Change, Maximize Success, Chronicle LLC, San Francisco, CA, 2004. 17. Print.
2. Potts, R., LaMarsh J., Change—The Basics. Master Change, Maximize Success, Chronicle LLC, San Francisco, CA, 2004. 37. Print.
3. Potts, R., LaMarsh J., Change—The Basics. Master Change, Maximize Success, Chronicle LLC, San Francisco, CA, 2004. 52–59. Print.

9

Project Risk Assessment

What follows is a change readiness checklist you can use as a preliminary assessment of your project's readiness to move forward from a change management perspective (Figure 9.1).

One extremely important part of change management is actually a very common project management tool, but it's often neglected or underwhelmed. Do a good job of creating a project charter and you will have taken a great step forward. Neglect it, and you put your entire project at risk.

Figure 9.2 is a project charter template that I found among the materials available on LeanMethods.com, a great source of free Lean Six Sigma information.

Once you have what you and your project champion believe to be a solid project charter in place, there is an important follow-up step that is often neglected—the completion of the project charter risk assessment.

Key Roles—Sponsor versus Champion?

The terms Sponsor and Champion are sometimes used synonymously. For purposes of this book, here is the distinction.

Sponsor

The senior executive who sponsors the overall Lean Six Sigma initiative.

Champion

Middle- or senior-level executive who sponsors a specific Lean Six Sigma project, ensuring that resources are available and cross-functional issues are resolved.

☑ Have you defined the change? Have you used a project charter (see Figure 9.2) to define the business gap? The problem statement? The objective? The scope? (What's in scope? What's not in scope?) The financial impact? The milestones and timeline? The project team and champion? See more details in the next section on project charter risk assessment (Figure 9.3).

☑ Have you identified the Targets? The Targets are the people who must change to make the project successful.

☑ Is there awareness of the need for this change? Are the Targets of the change aware of the need for change?

☑ Are you sure the Targets are aware of the need?

☑ Are you absolutely sure?

☑ Are the Targets' bosses (and their bosses' bosses) aware of the need for change? Are they supportive of the change?

☑ Do the Targets understand the business case for making the change? The WIIFTB, i.e., the What's in it for the business?

☑ Do you think the Targets also have a personal desire to support the change? Do the Targets understand the WIIFM?, i.e., the What's in it for me?

☑ If you sense there might be resistance to this project, have you asked for help?

FIGURE 9.1
Change readiness checklist. (Created by the author.)

Project Charter			
Project Title: xxx		Black Belt: xxx	
Business Gap x	Financial Impact $x		
	Milestones/Timeline: Define Tollgate Review: x Measure Tollgate Review: x Analyse Tollgate Review: x Improve Tollgate Review: x Control Tollgate Review: x	Scheduled	Actual x
Customer External: x Internal: x			
Defects and Metrics Defects: x Primary: x Secondary: x Consequential: x	Project Scope/Boundaries: Process Start: x Process Stop: x In Scope: x Out of Scope: x		
	Team: x		
Problem Statement x	Champion: x		
Objective Statement x			

FIGURE 9.2
Project charter template. (From http://www.leanmethods.com; Permission to use granted by Lean Methods (BMGI).)

Change Management Elements: a few examples

			Red/Yellow/Green	What are the Problem or Issues?
Leadership				
	The accountable sponsor/champion visibly champions the project.			
Charter				
	Business Gap/Problem Statement			
		The pain to be remedied or opportunity to be seized has been clearly defined.		
	Objective/Goal Statement			
		The specific results expected for this project are clearly described (qualitatively and quantitatively).		
	Business Impact/Finanical Impact			
		This change contributes to the accomplishment of an important buisness strategy, i.e., there is alignment.		
		There is a clear external threat or impetus driving this change.		
		The project "competes well" with other projects in the portfolio of influence of relevant senior leaders.		
		The business impact is quantified and compelling.		

FIGURE 9.3
Project charter risk assessment. (Created by the author.)

More on the Role of the Champion

If you ask people what a Champion does, they will quickly reply, "They remove roadblocks." Superficially, that is true. Champions should remove roadblocks. Champions need to be in a position to defuse any issues that may arise between a Black Belt and another person in the organization, particularly if the issue is with someone with a higher *formal* position in the company. The Champion should be the buffer that keeps a Black Belt out of a head-to-head confrontation with Managers, Vice Presidents, and Directors in the company, allowing Black Belts the freedom to focus on the problem, not engage in some inane territorial dispute. This is the most fundamental function of the Champion.

Source: http://www.isixsigma.com

SOME CHANGES ARE BAD CHANGES

Occasionally, you will run across a change that is, there's no other way to say it, a bad change. It might be a bad idea, or it might be a good idea, just ahead of its time. I've used resistance information gathered using organizational change management tools to explain and typically convince leadership that we should cancel or postpone that change project. Keep in mind, this is the rare exception and not the rule.

HOW TO PREVENT BAD PROJECTS

If a proper project selection and project prioritization process is used, this will greatly reduce the probability that a Black Belt will ever be assigned to a bad project. It's still possible, but much less likely.

KEY TAKEAWAYS

- You can use the change readiness checklist described in this chapter to conduct a preliminary assessment of your project's readiness to move forward, from a change management perspective.
- The project charter is an extremely important part of change management. It's actually a very common project management tool, but it's often neglected or underwhelmed. Don't make this mistake.
- You should take the time to assess the risks associated with the project charter carefully. The project charter risk assessment tool can help with this.

10

Some Basic Organizational Change Management Tools

I'm going to show you a few basic change management tools you might want to use on your projects. Let me start by suggesting that you should first engage your human resources team, or your resident organizational change management (OCM) expert, if you are lucky enough to have one. Please don't try to do them alone with these OCM tools, especially on your first projects.

STAKEHOLDER ANALYSIS TOOL

Stakeholder

> People who will be affected by the project or can influence it but who are not directly involved with doing the project work. Examples are managers affected by the project, process owners, people who work with the process under study, and internal departments that support the process, customers, suppliers… (e.g., the finance and accounting department).
>
> Source: https://www.isixsigma.com/dictionary/stakeholder

What Is the Tool Used For?

The stakeholder analysis tool, shown in Figure 10.1, helps you identify the people who can have the most positive, and the most negative, impact on your project.

This tool helps assess the positioning of stakeholders relative to change and commitment to the goals of the team, and it is used to identify and

Stakeholder Analysis

		A	×	B	×	C	=	D	
		Impact		Influence		Attitude		Score	
Stakeholder Name	Stakeholder Role in Organization	Impact of Project on Stakeholder (High = 3, Med = 2, Low = 1)		Stakeholder Level of Influence on Project's Success (High= 3, Med = 2, Low = 1)		Stakeholder Current Attitude Toward Project (Positive = +2, Neutral = +1, Negative = −2		Stakeholder Score (Range= −18 to +18)	Reasons for Resistance of Support
Jane Ryan	Operator	3	×	2	×	+1	=	6	Job will be easier
Fred Jones	Operator	3	×	2	×	+2	=	12	Job will be much easier
Jack Williams	Supervisor	3	×	3	×	−2	=	−18	New skills needed
Susan Brown	Supervisor	3	×	3	×	+1	=	9	Skills not an issue
			×		×		=		
			×		×		=		

FIGURE 10.1
Stakeholder analysis tool. (Created by the author.)

enlist support from stakeholders. It provides a visual means of identifying stakeholder support so that you can develop an action plan for your project.

The output of the tool is a representation of where the impacted people, both inside and outside the system, stand relative to the change.

How to Use the Tool?

Start by identifying the people who will be impacted most by the project. As you will recall, Jeanenne LaMarsh calls them the "Targets" of the change. Next, add the names of the people who will have the most influence over the project's success. Some of these influencers will also be "Targets" of the change, but others might be people in supporting, supervisory, or leadership roles.

For each person listed in the stakeholder analysis table, you assign an impact score ranging from 1 (low impact), to 2 (medium impact), to 3 (high impact). The impact score represents your estimate of the impact of the project on that person.

For each person listed in the table, you now assign an influence score ranging from 1 (low influence), to 2 (medium influence), to 3 (high influence). The Influence score represents the amount of influence you believe that person can have on the success (or failure) of the project.

I've found that the scoring process sometimes leads to remembering other people's names that should be added to the table.

For each person listed in the table, you now assign an attitude score ranging from −2 (negative attitude), +1 (neutral attitude), to +2 (positive attitude). The attitude score represents your current impression of how supportive a person is with regard to the project objectives. This is a hard one, as it calls for some guesswork on your part, based on what you've seen and heard. You might want to tap into some of the "F-word" stuff we discussed earlier … feelings. These scores often change over the course of the project.

Another way to think of the attitude rating is to consider the person's willingness to embrace the change and to make a cultural commitment to advocate for the change.

Next, you multiply impact times, influence times, and attitude scores to arrive at an overall score for each person.

$$\text{Stakeholder score} = \text{impact} * \text{influence} * \text{attitude}$$

If the math was done correctly, you should have scores ranging from −18 to +18.

Negative scores indicate people who might be resisting the project's objectives. You may want to focus your resistance mitigation efforts on them.

Note: Some stakeholder analysis tools include a "desired rating/stakeholder score" in addition to the "current rating/stakeholder score" and include an action plan designed to help make movement toward the desired target.

When to Use the Tool?

- Consider using this tool at the beginning of every project.
- You might also want to use it when you see possible misalignment of elements of the project charter among the sponsor/champion, the project team, and the Black Belt.
- Use it whenever the risk assessment seems (feels) incomplete.

What to Do with the Results?

- Identify areas of risk. They might be the larger negative stakeholder scores.
- Use findings from this report as inputs into the charter, communication plan, and the risk analysis.

RISK ANALYSIS TOOL

What Is the Tool Used For?

The risk analysis tool is based on the failure mode and effects analysis (FMEA) tool that is used in the Lean Six Sigma methodology (Figure 10.2). You probably have been, or will be, exposed to the FMEA approach and tool in your Black Belt training. The risk analysis tool is used to identify potential risks to the project, define how they could impact the project (scope, cost, time), define the probability of occurrence, estimate the impact relative to the project, and calculate a priority rating by multiplying probability of occurrence by the estimated impact. A risk response is identified for each risk and assigned to an owner who is charged with ensuring that the risk response is properly executed.

How to Use the Tool?

Start by identifying the risks that might impact the project. Next define the type of risk. Will it potentially impact scope, cost, and/or time? Next assess the probability of this risk occurring. If there is no perceived risk,

Risk analysis

Risk #	Description of the Risk	Impact Type	Probability of Occurrence	×	Impact Rating	=	Priority Rating	Risk Response		Owner
		Scope Cost Time	1 = None 2 = Low 3 = Medium 4 = High		1 = None 2 = Little 3 = Moderate 4 = Heavy		(Probability × Impact)	Accept Avoid Reduce Probability Minimize Impact	Identify and Describe Risk Response	
1	Some functional leaders will resist the change.	Cost and time	3	×	4	=	12	Reduce probability	Engage OCM expert	John S.
2				×		=				
3				×		=				
4				×		=				
5				×		=				
6				×		=				

FIGURE 10.2
Risk analysis tool. (Created by the author.)

assign a 1 (although I wonder why you included it on the list in the first place). If the risk is low, assign a "2"; if the risk is medium, assign a "3"; and if the risk is perceived to be high, assign a "4."

Next, assess the relative impact to the project should the risk occur. If there would be no impact, assign a "1"; if the risk is low, assign a "2"; if the risk is medium, assign a "3"; and if the risk is perceived to be high, assign a "4."

Now, multiply probability of occurrence by impact rating to arrive at the priority *rating for each risk.*

Next, consider the risk response for each risk. Should we just accept it? Avoid it? Work to reduce the probability of occurrence? Or minimize the impact?

Next, identify and describe the risk response along with the owner of said response.

When to Use the Tool?

- Consider using this tool at the beginning of every project.
- Use it whenever the risk assessment seems (feels) incomplete.

What to Do with the Results?

- Identify the areas of greatest risk. The highest risk priority ratings.
- Make sure the risk responses, i.e., mitigation plans, are satisfactorily executed. If not, notify your project sponsor/champion immediately.

COMMUNICATION PLAN

What Is the Tool Used For?

I've found that OCM is mostly about communications, and most people, and most companies tend to under communicate about important planned changes. The communications plan helps to define a structured delivery of the key messages to the right audiences, delivered by the right people, using the right medium, at the right time and frequency (Figure 10.3).

Communication plan

Audience	Media	Purpose	Key Message	Owner	Frequency	Status
Senior Leadership Team	Face to Face Meetings w CEO	To make them Aware of the need for change	We have a problem	CEO	Repeat for top two levels of leadership	Meeting #1 - Feb 12 Meeting #2–4 - Feb 13

FIGURE 10.3
Communication plan. (Created by the author.)

How to Use the Tool?

The input for the communication plan comes from many sources. The resistance identified using other tools often results in the need for coordinated communication. At the start, most of the communication is about the "Why"—why change? The messages are often delivered by leadership to various audiences to build awareness of the change, and to start developing some personal desire for people to participate in the change.

Audience groups might include the senior leadership team, the Targets of the change, or the supervisors of the Targets, with each audience getting a slightly different message to help it understand the Why, the business case, and the WIIFM (What's in it for me?).

When to use the tool?

The easy answer is, from the beginning of the project to the very end, and everywhere in between. I have not seen a case where there has been too much good communication. I have seen instances of too much bad, poorly crafted communication, which did not help.

What to do with the results?

Just do it! Execute the plan, on time and with the designated "owner"—no substitutes: e.g., no last minute stand-ins for the CEO. This is important!

Some lessons learned regarding communication:

- Communicate, communicate, communicate—most communicate too little
- Communications should include two-way discussions

Some Basic Organizational Change Management Tools • 65

- Include supervisors
- Communicate in many ways—face-to-face, in "town hall" meetings, via email, etc.

LEVERS YOU CAN PULL

With consideration given to the OCM tools shown in this book, and the myriad of similar OCM tools available elsewhere, I've found three primary actions that tend to result from all of them. You might think of them as the levers that can be pulled...

The word "tools" in OCM is a bit of a misnomer, because most of them are not like a hammer or a screwdriver. Most are not designed to DO something. Most are designed to help you PREPARE to do something (Figure 10.4).

There are three main "levers" you can pull to manage resistance and increase the probability that your change will be successful and sustained. You can use

1. Communication
2. Incentives (positive and negative, i.e., carrot and stick)
3. Training and development

In my experience, here's how the distribution of "lever pulling" typically goes (Figure 10.5). Please keep in mind that the use of the "levers" varies widely from project to project. This is just my estimate after being involved in many change projects over the years.

Change management is typically

FIGURE 10.4
Pulling an OCM lever. (Created by the author.)

66 • *The Intersection of Change Management and Lean Six Sigma*

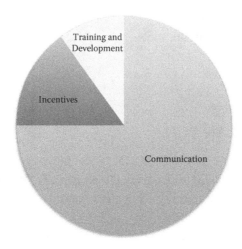

FIGURE 10.5
Change management—When viewed from 50,000 ft. (Created by the author.)

Mostly about communication
Next, about incentives ("carrots and sticks")
And then, about training and development

Communication

Careful now. It's not just about how much to use the various change management tools and levers, it's also about *when*. The timing is critical. When working on a change, I've seen many project teams lead with training, which is almost always a bad idea. It's not a good idea to train people until after the participants have acquired awareness and desire. If the training is done earlier, it's likely to be a waste of time and money. And, just try getting the Targets back in the class again a few months later for a repeat performance, when they're finally ready and willing to be trained. Their bosses may push back on this.

The solution is to train people once: when they are ready, engaged, have awareness and desire, and are receptive to being trained.

Incentives

Incentives might also be called reinforcements. These are so-called "carrots and sticks" that can be designed and implemented by the leadership and management team. I've found these to be less used on Lean Six Sigma Black Belt projects, but if there is a perceived need, you should involve your project sponsor/champion and engage your friends from human resources. Don't try to take on incentives alone.

Lisa Riegel, Ph.D, recently reminded me that, "we don't always need money incentives. Sometimes positive reinforcement can really help… and that can simply be communicating that a person did the right thing and is a valuable member of a team."

Training and Development

A simple test that I've seen applied to help identify the need for training and development is to ask the question, "Could <insert a Target's name> perform this task if their life depended on it?" In other words, is it a matter of will or skill?

If it is determined to be a matter of will (or lack thereof), we need to determine the problem. What is the obstacle? Is there a lack of motivation? A lack of awareness? No desire? Is it resistance?

If it is a matter of skill (a lack thereof), we need to determine the best approach to address the competency problem. Look for skill gaps—assess, then address. Possibly coaching, a job aid? Formal training? If none of these approaches work, then maybe the person needs to be shifted to a different job.

Sidenote: A word about training: There's something called the Kirkpatrick Model that has been around for years. It's used to evaluate the effectiveness of training.

How many of you have attended a training class and at the end you're asked to complete what is the equivalent of a smiley face evaluation form? If you liked the environment, trainer, materials, etc. you'd circle the ☺. If you didn't like it, you'd circle the ☹. This is Level 1 learning, which involves assessing a participant's *reaction* to the training.

Level 2 learning measures the degree to which participants *acquire* the intended knowledge skills, attitude, confidence, and commitment based on their participation in the training. A pretest, posttest approach applied to a training session is one way to assess Level 2 learning.

Level 3 learning involves the degree to which participants *apply* what they learned during training when they are back on the job.

Level 4 learning involves the degree to which targeted *outcomes* are achieved as a result of the training.

In summary, the Kirkpatrick Model includes four steps:

Level 1: Reaction
Level 2: Learning
Level 3: Behavior
Level 4: Results[1]

More about the Kirkpatrick Model can be found at www.kirkpatrickpartners.com.

KEY TAKEAWAYS

The stakeholder analysis tool helps you identify the people who can have the most positive, and the most negative, impact on your project.

- The *risk analysis tool* is used to identify potential risks to the project, define how they could impact the project (Scope, Cost, Time), define the probability of occurrence, estimate the impact to the project, and calculate an overall priority rating.
- The *communications plan* helps to define a structured delivery of the key messages to the right audiences, delivered by the right people, using the right medium, at the right time and frequency.
- There are three main "levers" you can pull to manage resistance and increase the probability that your change will be successful and sustained.
- You can use:
  - Communication
  - Incentives (positive and negative, i.e., carrot and stick)
  - Training and development

REFERENCE

1. Kirkpatrick Partners, The One and Only Kirkpatrick Company˙. The Kirkpatrick Model. *The Kirkpatrick Model*. N.p., n.d., March 19, 2017. http://www.kirkpatrickpartners.com.

11

When to Ask for Help

There comes a time in every Black Belt's career that they find themselves in "over their head" (from an organizational change management [OCM] perspective) on a project. I've used the image of a beach and seashore to help portray some situations with regard to the need to ask for help in dealing with resistance.

The depth of the water in Figure 11.1 represents the amount of resistance to change on a project. If you find yourself as Black Belt/project leader getting into "deep water" with regard to resistance, seek out help from an OCM expert. If no expertise exists at your site, call someone. I recommend you start your search for OCM expertise in your human resources department. People there probably know someone who can help.

In the situation shown on the left side of Figure 11.1, the Black Belt is wading in to shallow water. Things are fine. The sun is shining. The waves are lapping at his/her feet. There is minimal resistance to his/her project. In this case, good intuition regarding change management may be enough. More on intuition later.

In the situation shown in the middle of Figure 11.1, the Black Belt is swimming in deeper water. He/she is facing some mild resistance on the project, and the use of some simple OCM tools might suffice.

In the situation shown on the right side of Figure 11.1, the Black Belt is in deep water, facing much resistance. Sharks appear in the water. It's time to seek help from an OCM expert.

Staying with the swimming analogy for a minute, I'd like to make an observation regarding Lean Six Sigma training delivered as an initial step in what some call a Lean Transformation.

I've attended Lean training sessions, and I look around the room knowing that 95% of the people in the room don't get it. They're no longer listening. Why is this? The training is often delivered in a "cookie cutter"

70 • *The Intersection of Change Management and Lean Six Sigma*

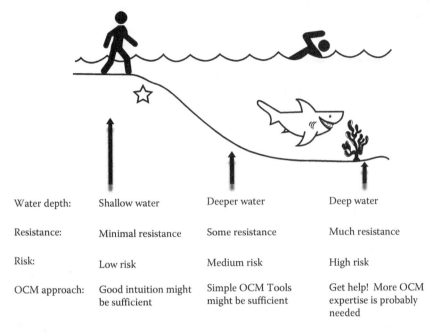

Water depth:	Shallow water	Deeper water	Deep water
Resistance:	Minimal resistance	Some resistance	Much resistance
Risk:	Low risk	Medium risk	High risk
OCM approach:	Good intuition might be sufficient	Simple OCM Tools might be sufficient	Get help! More OCM expertise is probably needed

FIGURE 11.1
When to ask for help. (Created by the author.)

approach to disseminate information across the organization. I think of this as the equivalent of a senior livesaving training class, where some fairly high-level lifesaving approaches are being taught to people who may not yet even know how to swim. A few people in the room get it and leave with the plan to go do something with the training, while the many who don't know how to swim listen and try to connect for a while, but then ultimately check out. These people need to be taught how to float. My point in all of this is that the training needs to be tailored to the audience, which is not easy to do when you're in the mass delivery mode.

OCM INTUITION

I've noticed that some people seem to have GREAT intuition about OCM. They naturally communicate with people across the organization—up, down, and across. They tend to exhibit the qualities of a great leader: respect and humility. They seem to spend a lot of time involved with the three C's—communication, cooperation, and collaboration. I'd put these people on the right side of the distribution in Figure 11.2.

When to Ask for Help • 71

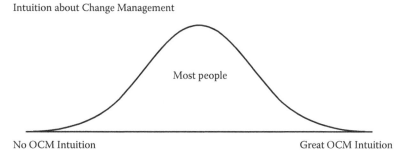

FIGURE 11.2
OCM intuition distribution. (Created by the author.)

I've also noticed that other people fall on the left side of the distribution in Figure 11.2. They often would not score high marks on things like communication, cooperation, collaboration, respect, or humility. Don't waste much time on these people. With regard to OCM, they probably don't get it, don't believe in it, and won't support it.

KEY TAKEAWAYS

- If you find yourself as Black Belt / project leader getting into "deep water" with regard to resistance, seek out help from an OCM expert.
- I recommend you start your search for OCM expertise in your human resources department. The people there probably know someone who can help.
- Some people seem to have GREAT intuition about OCM. They naturally communicate with people across the organization—up, down, and across. These people are typically great to work with on change projects.
- Other people do not have very good intuition about OCM. These people may not receive high marks on things like communication, cooperation, and collaboration. And, they may not support the notion of earning respect from their people or demonstrating humility in the workplace. Don't be surprised if they don't see the need for OCM and won't support it.
- If you encounter these nonsupportive people, you should seek out others in management with attributes that support your change work.

12

The Intersection of Organizational Change Management and Lean Six Sigma

Years ago, I was trained and certified as a Black Belt. My lead instructor was a man by the name of Dr. A. Blanton Godfrey. Blan has spent a lifetime improving the world, through his roles in industry and in academia. During my time learning from Blan, he talked extensively about the need for change management, but the training and certification program at that time did not include training in organizational change management (OCM) tools. Blan later rectified this situation.

Dr. Godfrey's impressive resume could easily fill a chapter in this book. Let me summarize his career by telling you that Dr. Godfrey is listed along with Deming, Juran, Shingo, Shewhart, Feigenbaum, Ishikawa, Taguchi, Crosby, Goldratt, Ohno, and only 40 others on www.qualitygurus.com.

Today, some Lean Six Sigma (LSS) training programs include change management modules, but few include extensive training in this area. It's more than just the tools. Would you allow a 16-year-old kid, equipped with a new set of power tools, to remodel your house? Probably not, unless that kid had extensive experience employing those tools successfully.

In your LSS training, you learned (or you will learn) that a five-step process called DMAIC is used on some projects to address a problem.

D = define
M = measure
A = analyze
I = improve
C = control

74 • *The Intersection of Change Management and Lean Six Sigma*

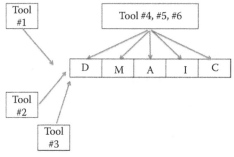

Tool #
1 Change readiness checklist (Figure 9.1)
2 Project charter template (Figure 9.2)
3 Project charter risk assessment (Figure 9.3)
4 Stakeholder analysis (Figure 10.1)
5 Risk analysis (Figure 10.2)
6 Communication plan (Figure 10.3)

FIGURE 12.1
Mapping of OCM tools to the DMAIC steps. (Created by the author.)

The mapping diagram shows where some simple OCM tools can be used throughout the DMAIC process to help identify and mitigate resistance to change (Figure 12.1).

Note that the application of OCM is not limited to DMAIC. It's equally valuable when used in conjunction with the DMADV steps and the Kaizen process.

The intersection of LSS and OCM refers to the intertwining of these two complimentary approaches and methodologies. As pointed out by Ms. Pandya in Chapter 1, "… you cannot have Continuous Improvement without Change Management." Michele Quinn's quote in Chapter 8 helped us understand that OCM aligns perfectly with LSS in that they are both *structured processes* that, when used properly, *deliver organizational results*.

When the intersection of OCM and LSS is done correctly, the Black Belt uses OCM approaches and tools as part of the DMAIC process. (Figure 12.2)

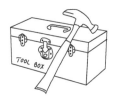

FIGURE 12.2
A partial solution. A tool box with PM, LSS, and OCM tools. (Created by the author.)

The Belt no longer thinks of these as change management tools. They are simply additional tools in the Black Belt LSS tool kit.

A NOTE TO THE LSS PROGRAM/DEPLOYMENT MANAGER

Along with blending or sprinkling OCM tools into the Black Belt toolbox and the DMAIC steps, there's also another intersection between these two methodologies that occurs at the program or initiative level. In this case, in addition to using OCM tools to manage resistance to individual projects, you may also find the need to employ OCM tools to help reduce resistance to the overall LSS deployment (or redeployment).

As of this writing, LSS has been around for decades. Many organizations are in their 10th attempt at "implementing Lean." Outside the scope of this book is the need to find the REAL root causes of the repeated failures to implement (and sustain) Lean. See my book on Hoshin Kanri for some thoughts on this matter.

Figure 12.3 illustrates that the successful Black Belt:

- Has Black Belt competencies
- Has skills in project management (PM), LSS, OCM
- Holds a tool box filled with PM, LSS, and OCM tools
- Using a Shingo-like approach based on respect and humility

FIGURE 12.3
The successful Black Belt. It's not just about the tools. (Created by the author.)

KEY TAKEAWAYS

- Most organizations today are in their 10th LSS deployment or Lean Transformation, and some of the training programs include change management modules.
- Few organizations go beyond the "module" training of OCM, so most Black Belts leave certification armed with only a few change management tools.
- OCM is much more than a few simple change management tools.
- And, becoming a successful Black Belt involves much more than learning a set of PM, LSS, and OCM tools. It's not just *what* you know, it's *how* you apply it... how you interact with people every day.

13

People Are *Different*

Under the laws enforced by EEOC, it is illegal to discriminate against an applicant or employee based on race, color, religion, sex (including gender identity, sexual orientation, and pregnancy), national origin, age (40 or older), disability, or genetic information.[1]

When I say people are different, I'm not talking about any of the aforementioned traits. I'm referring to differences in workplace personalities and thinking styles. And some, including this author, believe these differences are *measureable* and are actually *important* to help improve the performance of the organization.

When I've conducted personality or thinking style assessments in the past, I've found that many Black Belts tend to have scores that fall on the left side of the DISC model (see Figure 13.1). In other words, they tend to be more task than people oriented. It seems to me (no data here) that they often tend to like and rely upon facts and data more than the people on the right side of the DISC model. But, I've also known many highly successful Black Belts with scores that placed them on the right side of the model.

The thing to remember is that the world's population, and possibly the teams with whom you're working, have workplace personalities and thinking styles that are equally distributed around the four quadrants of the DISC model shown in Figure 13.1.

Important: Just because you like facts and data doesn't mean that the people on your project team, and (most importantly) the Targets of your change, think and feel the same way! Communication is only effective if you present people with information in a way that they will be likely to hear it. And, I mean REALLY hear it. My wife Susan, for example, is not a huge facts and data person. However, much to my chagrin, I've come to realize that when she wants something—and ultimately gets it—she presents me with a business case. She has learned to mitigate my resistance by speaking my language.

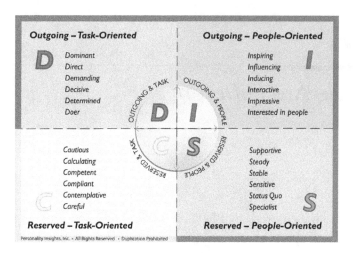

FIGURE 13.1
The DISC model. (The intellectual content is provided with permission by Dr. Robert A. Rohm & Personality Insights, Inc., www.personalityinsights.com. All rights reserved. Duplication is prohibited.)

Another way people can be different is in terms of their ability to deal with change. Years ago, I was certified in the CENTACS WorkPlace Big Five methodology, and I learned that people's workplace personality can be assessed across what are called five supertraits.

INTRODUCTION TO THE WORKPLACE BIG FIVE PROFILE

In 2001, the Center for Applied Cognitive Studies (founded by Pierce and Jane Howard in Charlotte, North Carolina, and now called the Paradigm Personality Labs) introduced the WorkPlace Big Five Profile, a survey with workplace-oriented language. The current version of the Profile measures 5 supertraits and 23 subtraits.

The five supertraits that describe work-related behavior are as follows:

Need for stability (N)
Extraversion (E)
Originality (O)
Accommodation (A)
Consolidation (C)

What follows further defines each supertrait and identifies the subtraits found within each supertrait.[2]

NEED FOR STABILITY (N)

Explains how people at work respond to and handle stressful situations, a critical aspect of today's successful and effective work environment.

1. Worry
2. Intensity
3. Interpretation
4. Rebound time

EXTRAVERSION (E)

Defines how people at work tolerate and deal with sensory bombardment or the lack of it. Situations would include when people work alone at home or settings and sensory experiences such as a three-day senior management off-site strategy meeting.

1. Warmth
2. Sociability
3. Activity mode
4. Taking charge
5. Trust of others
6. Tact

ORIGINALITY (O)

Illustrates how open and accepting people at work are to new experiences, ideas, and change.

1. Imagination
2. Complexity
3. Change
4. Scope

ACCOMMODATION (A)

Measures how easily or to what degree of difficulty people defer to others—this supertrait relates directly to power and how to use it effectively.

1. Others' needs
2. Agreement
3. Humility
4. Reserve

CONSOLIDATION (C)

Explains the degree to which people at work focus on their work, goal accomplishment, and need for achievement and success.

1. Perfectionism
2. Organization
3. Drive
4. Concentration
5. Methodicalness

Let's talk about just two of these supertraits—(N) need for stability and (O) openness.

Important: If a person, maybe it's one of your Targets for change, is assessed as having a high need for stability, i.e., a high "N" score, they might be more likely to resist the change you are proposing. Likewise, if another person involved with the project has a high "O" score, an openness to consider new things, they might be more likely to embrace the change. On an important change project, it might be good to know things like this before you start your plunge into the Delta State.

In addition to the DISC model and the CentACS WorkPlace Big Five methodology, there are other tools available to identify important differences in people's workplace personalities and thinking styles. One such tool is called the Change Style Indicator.

The Change Style Indicator, offered by Multi-Health Systems, Inc., is an assessment instrument designed to measure an individual's preferred style in approaching change and in addressing situations involving change.[3]

Three primary styles exist on a continuum: conserver—pragmatist—originator.

More info on this tool can be found in Appendix E.

Sidenote: Change management includes attending to three component systems: technical, administrative, and people. At least, that is one way of looking at it. The change manager can assume that all screws in one batch are interchangeable (technical system). Similarly, they can assume that all dollars are equivalent (administrative system). But they cannot assume that all workers are interchangeable (people system).

People vary according to traits, mental abilities, experiences, physical characteristics, and values. They are not interchangeable like screws and dollars. Perhaps for a temporary period, you can put a worker in a position for which they lack the ideal qualities, as in putting a more big-picture person in a proofreading role. They can adapt to these unnatural tasks for a period of time, but eventually they will feel a strong urge to revert to their set point of that trait, ability, value, etc.

As a change manager, you need to know whether you are assigning someone to a role for which they are a natural fit, or whether there are areas of misfit that you will need to allow for. If a worker continues for too long that involves a significant misfit, you are risking injury, quality deficits, job dissatisfaction, stress, and, eventually, burnout.

—**Pierce J. Howard, Ph.D.**
Co-Founder and Chief Innovation Officer
Paradigm Personality Labs

Remember the story about Ph.D. Stan back in Chapter 5? He was a real person (given a fictitious name for this book), and he was one of the biggest lovers of facts and data I've ever met. I'll bet you know people like this. Maybe *you're* one of them? I'll bet you also know people who use the F-word (feelings) in almost every meeting at work. It's possible, but chances are that the person who talks about feelings when making business decisions is *not* also a facts and data lover.

People are different. You can probably use facts and data, charts and graphs, to help some people gain awareness and personal desire to make a change. With other people, you can be armed with the best data and the most convincing histogram, and it won't budge them an inch. You'll need to find another approach to convince some of the "feelers" to have a personal desire to make an important change.

CHANGE PROPENSITY

I'll caution you ahead of time—I have no facts and data to support the "bell curves" you'll see in this chapter—only a lifetime of observations. Dr. Pierce at Paradigm Personality Labs (www.CentACS.com, soon to be www.ParadigmPersonality.com) could probably provide some statistical data in support of my supposition, if queried. My observation has been that a few people seem to LOVE almost every change they encounter, while others absolutely HATE almost any change. Most of us can be found somewhere in between. It's situational. It depends on the change. It depends on what else is going on in our life (personal and business) at the time (Figure 13.2).

PEOPLE AND FACTS AND DATA

I've also observed that a few people LOVE facts and data. They live and breathe it. While a few others don't want to ever see a number or a graph. For them, it's all about feelings and intuition. Keep this in mind when designing your communication plans, your incentives, and your training and development plans (Figure 13.3).

The bottom line is that people *are* different, and in important ways that can really matter at work.

Some people's DISC scores tend to fall on the left side of the DISC profile, and they are task oriented and might be more inclined to consider (and maybe even love) facts and data. While some people's DISC scores fall on the right side of the DISC profile and they tend to think more about people and feelings, and possibly less about facts and data, when making decisions at work.

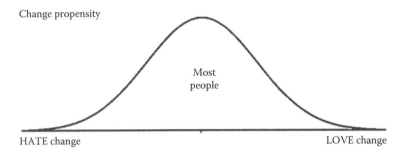

FIGURE 13.2
A hypothetical distribution: Propensity to like change. (Created by the author.)

As explained during the WorkPlace Big Five discussion, some people are inclined, whether by nature (their genetic makeup) or nurture (their environment), to need more stability (less stressful situations), while others tend to thrive in less stable, even chaotic situations. Some are more open to change and new experiences, while others will almost always opt for sticking with the status quo.

I've found that it's often good to understand the workplace personalities and thinking styles of those people with whom you're working on a project, especially when dealing with a change that could face significant resistance.

It's better to know it and not need it, than to need it and not know it.

PEOPLE ANALYTICS

Shown in Figure 13.4 is what I'm calling People Analytics, a bridge between the people side of change and the hard data used by the Lean Six Sigma Black Belt. Tools such as CENTACS' WorkPlace Big Five and

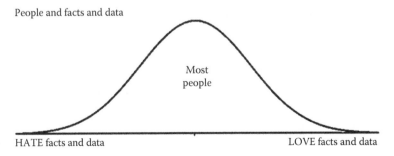

FIGURE 13.3
A hypothetical distribution: Propensity to like facts and data. (Created by the author.)

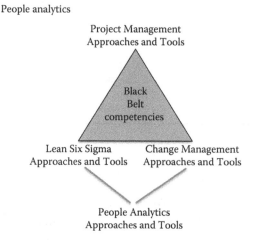

FIGURE 13.4
People Analytics: A bridge between the "People Stuff" and hard data. (Created by the author.)

Syndio Social make use of data, statistics, and other analytical tools to derive information about people's behavior and workplace personalities. This helps remove some, but not all, of the subjectivity from the analysis.

KEY TAKEAWAYS

- There are differences in workplace personalities and thinking styles that matter at work.
- These differences are *measureable* and are actually *important* to help improve the performance of the organization.
- The teams with whom you're working have workplace personalities and thinking styles that are typically equally distributed around the four quadrants of the DISC model. That means that three-fourth of the people on your project team may not think the same way you do!
- People can also be different in terms of their ability to deal with change. The WorkPlace Big Five methodology helps with this by assessing a person's workplace personality across what are called five supertraits.
- Some people seem to LOVE almost every change they encounter, while others absolutely HATE almost any change. Most of us can

be found somewhere in between. It's situational. It depends on the change. It depends on what else is going on in our life (personal and business) at the time.
- Some people seem to LOVE facts and data. They live and breathe them. While a few others don't want to ever see a number or a graph. For them, it's all about feelings and intuition. You can probably use facts and data, charts and graphs to help these people gain awareness and personal desire to make a change. With others, you can be armed with the best data and the most convincing histogram, and it won't budge them an inch.

REFERENCES

1. Prohibited Employment Policies/Practices. *U.S. Equal Employment Opportunity Commission*. N.p., n.d., March 19, 2017. http://www.eeoc.gov/laws/practices.
2. Howard, P., and J. Howard. Why the Big Five? *Centacs*. N.p., n.d., March 19, 2017. Centacs.com is rebranding to become ParadigmPersonality.com shortly after publishing of this book. http://www.centacs.com.
3. Change Style Indicator. *MHS Assessments*. N.p., n.d. http://www.tab.mhs.com.

14

The Final Word

Question: What do you call the person who finished last in their class in medical school?
Answer: Doctor
OK, I threw that one in just for fun.

Question: Whom do you want operating on you, a newly trained M.D. or a doctor with many successful operations under her/his belt?
Answer: I want the operation performed by an experienced doctor who made any early career mistakes on someone else and *learned* from them.

The same is true with regard to Black Belts and change management. The best Black Belts tend to be those who have worked several projects, made some mistakes, learned from each of them, and then took corrective action to avoid repeating them.

I'm glad you took the time to read this book explaining the need for change management on Lean Six Sigma change projects. Are you now completely ready to employ change management approaches and tools on your Lean Six Sigma projects? Probably not.

Does the medical doctor in training go off and perform surgery after reading the first Intro to Anatomy book? I hope not!

You, and the medical doctor in training, have more *learning* and *practicing* to do.

Now, go forth and learn and learn more ... maybe even become certified in a change management methodology ... and then do good work!

KEY TAKEAWAYS

- As Ellen Domb recently reminded me, this book is really all about WASTE (a Lean term)! It is wasteful to have a team do a great job of improving a process, then have the new process not be adopted by the organization due to lack of understanding of the human side of changing to the new process.
- I encourage you to learn a change management approach and then use it to help minimize the resistance (and WASTE) that is otherwise inherent in your continuous improvement activities.

Section III

15

Interviews with Experts

What follows is the result of numerous interviews with Change Management experts—practitioners in industry, consultants, authors, Master Black Belts (MBBs), and professors. Many of these experts work(ed) on the ground at companies like Danaher, Motorola, Caterpillar, Shell Oil, Bank of America, Wiremold, Wells Fargo, Coca-Cola, DuPont, Lowes, Duke Energy, Dana, and many others. I've scoured through countless pages of notes from my interviews and have included some of the most insightful responses about successes and failures with managing change. I hope you will find their comments as informative as did I.

As you read the experts' comments, you will see that they are *not* all completely aligned with each other, and they certainly do not all agree with me. After working at the intersection of change management and Lean Six Sigma for many years, I have learned that there is not one "right" way to do this stuff. I know some approaches and tools, but I'm always learning new things.

Sidenote: Over 100 people volunteered to read and comment on this book when it was still in its draft form. People seemed to either LOVE this chapter or they HATED it. There was not much middle ground. The haters told me that I should have edited this chapter and removed all of the "outlier" comments, i.e., I should have made the Q&A section agree with what I wrote in the front of the book. Others said that I should have summarized the interviewees' thoughts to save readers from taking the time to read all of them. In the end, I left the chapter format alone. I like the fact that there is not complete agreement on this subject. It confirms my belief that there is more than one way to do this stuff. As Taylor Swift would say, "Haters gonna hate" ☺

THE INTERVIEW QUESTIONS

Q1. What is your personal experience with organizational change management?

Q2. What is your personal experience with Lean Six Sigma?

Q3. How would you explain the importance of change management to a Lean Six Sigma black belt (BB)?

Q4. What is change management, in your own words?

Q5. What three words would you use to describe change management?

Q6. Why use change management?

Q7. How have you seen change management used in conjunction with Lean Six Sigma? Was it successful?

Q8. What do you think the typical Lean Six Sigma BB needs to know about change management? How should it be taught to the belts?

Q9. What are the critical success factors, i.e., the secrets to success, in using change management in conjunction with Lean Six Sigma?

Q10. Lessons learned in using change management (with or without Lean Six Sigma)?

Q11. What can go wrong when using change management?

Q12. What are the dos and don'ts with regard to change management?

Q13. Do you have a favorite change management methodology? If yes, please explain.

Q14. Key metrics? How do you measure the effectiveness of change management?

Q15. Please describe the biggest mistake you've made (or seen being made) with regard to employing change management.

Q16. Do all Lean Six Sigma projects need change management? If not, how do you know when it's needed?

Q17. How does a BB (the project leader) know when it's time to ask for help with a change management problem?

Q18. How can you measure the level of resistance to a project?

Q19. What is the project champion's role (some call it the sponsor) with regard to change management?

Q20. Are gaining adoption and eliminating resistance the same thing? If not, please explain the difference?

Q21. With regard to reducing resistance, how would you address a group of "thinkers," i.e., highly analytical, data-oriented people?

How would you address a group of "feelers," i.e., people who are not interested in facts and data?

Q22. Do organizations with excellent employee engagement still need change management?

Mini-Biographies of the Interviewees can be found in Chapter 16.

Q1. What is your personal experience with organizational change management?

Tom Cluley: organizational change management is at the heart of my company's approach to strategy execution, leveraging Lean and Six Sigma continuous improvement methodologies. I have training and assessment tools developed on many methods, associated with personality analysis, team effectiveness, effective communications, change management, leadership, and dealing with difficult people.

Bob Dodge: I am certified in LaMarsh as a change manager, change master, change management facilitator, methodology master, and also hold certification in Prosci. I have 17 years of practical experience with my own company, implementing improvements for my clients, and 3 years with LaMarsh clients, facilitating, assessing, coaching, and consulting.

David Ringel: I have 20 years of experience in the private industry as a process owner (i.e., shift manager, plant manager, operations manager, director of operations). In these positions, I was more able to direct change and manage it from that perspective. I also have 12 years of experience as a change agent/performance improver (i.e., director of Lean manufacturing in [the] private industry, 10 years as a performance improvement consultant).

Joe Beggiani: I have used organization change management for numerous projects, from rolling out new systems to training development to standing up a new organization. I was trained and certified in the LaMarsh Change Master Program in 2005.

Lynn Doupsas: I started my career in the organization and human performance practice of Accenture. I completed change management coursework and am also certified in MBTI, Firo-B, and NBI. I worked on and off for the last decade as a change manager and coach for Fortune 500 companies, both internally and as a consultant.

Scott Leek: I advised and consulted numerous leadership teams and belts on change management for more than 25 years. I trained on the subject as a consultant to GE Capital and later as an employee. I also taught the subject to clients and advised leadership teams on large-scale deployments.

Gary Bradt: I'm a licensed clinical psychologist in the state of North Carolina since 1987 and have been in the business of leadership consulting since 1987. I've had my own business since 1993. I do a lot of individual coaching with C-suite level executives who lead and manage change. I've also written books and delivered keynotes about leading through change.

Sarah Carleton: I attended a 1-week train the trainer (TTT) course in change management. I was certified as a qualified instructor and have taught many classes. I am also a certified MBB. I am a consultant who provides Lean and Six Sigma training with a focus on integral theory (based on Ken Wilber's works).

Roxanne Britland: My first introduction to change management was while I was running the International Society for Six Sigma Professionals (ISSSP), and I was running what we called Leadership Circles and MBB Circles, which were designed for one-hour to half-day sessions on a particular topic. The leadership level included champions, managers, and people who had belts reporting to them, with whom the topic of change management was very relevant.

With the MBB Circle, there was a big need to introduce the soft side of business into their more technical world and community, so I engaged a relationship with Lamarsh & Associates. I brought them into our event to run seminars on change management. That's how I was introduced to change management. Since then, I've worked with the US government and am currently bringing managing change principles and techniques into everything that we do to support organizational development (OD).

Darryl Bonadio: I have been involved in organizational change activities for the past 10–12 years. I was formally trained and certified in Accelerated Implementation Methodology (AIM), which is a change methodology approach. I have been a Six Sigma MBB since 2010, which also allotted me opportunities to influence organizational change.

Don Linsenmann: I led the Enterprise Process Management Office for DuPont.

Bob Von Der Linn: I am experienced with GE's Change Acceleration Process (CAP) and am Prosci/ADKAR certified. I'm also widely familiar with Kotter, Bridges, and others.

Marc Zaban Fogel: I started doing business development for a Lean Six Sigma consulting company based in China in 2002, and also started doing program management and sales for new projects in China at that time. We followed the John Kotter road map. I moved back to the United States doing the same in 2003. I've been helping companies use this road map for general projects as full-time work since 2008.

Beth Cudney: I was part of the initial wave of BBs that went through training on Six Sigma for our organization. My site had five people go through the Six Sigma BB training, and we led the change initiative in our organization.

Currin Cooper: I have been a project and change manager for almost 20 years.

PMI: Project Management Professional (PMP) since January 2007. Six Sigma Green Belt (GB) with DFSS (Design for Six Sigma) since 2006. I received change management certification with Acuity Institute in 2017. I am also a certified change management professional (CCMP) and a member of the Association of Change Management Professionals (ACMP).

I have been working as an organizational change management consultant since 2012 [at] Queens University of Charlotte, McColl School of Business, MS in Organization Development program.

Mike Ensby: I previously taught OB/OD courses at the undergraduate and graduate level and have consulted in this area for over 20 years, in a wide range of industries/sectors.

Patra Madden: I am a Prosci-certified practitioner, advanced practitioner, Enterprise Change Management Boot Camp, and Prosci TTT (lead trainer for a corporation of ~7000 people).

Scott McAllister: I've been involved in initiating organizational change through the last 15 or 20 years, both inside of organizations and as a consultant. I've been through Prosci's certification program and TTT program. I also do a lot of teaching in my current role. I completed the BMGI 5-day change leadership program as part of the MBB development process.

Abdiel Salas: I hold an MBB certification (SBTI, Inc.), am a certified Lean manager (Ohio State University and Productivity Inc.), and am a member of [the] Operational Excellence Committee for Emerson Latin America. I have been responsible for deployment of Lean Six Sigma programs in two companies.

Rick Rothermel: I have been a practitioner, consultant, and coach for more than 25 years. I hold multiple certifications in the LaMarsh Global Managed Change Methodology, as well as Prosci.

Michele Quinn: I am a trainer, facilitator, and coach in the area of organizational change. I'm certified as a Prosci master instructor and TTT.

Jessica Bronzert: I was Director Change Execution for Lowe's Companies prior to going out on my own as an executive coach and change management consultant. While at Lowe's, I helped transform the organization and culture to meet changing consumer demands. My primary change background is in the Conner methodology.

Siobhan Pandya: During my previous role at Shell, I was project manager for Shell's E2E Customer Experience project, which required the transition of key sales support activities (disputes, pricing, and customer set and amend) to customer service centers in Manila and Cape Town. This entire project was about organizational change management, from the design of the future organization to execution through communication, training, team working, counseling, etc. In addition, I have attended several internal Effective Change Management courses at Shell.

Q2. What is your personal experience with Lean Six Sigma?

Tom Cluley: I was trained by Shingijutsu in Lean while a member of the leadership team within Wiremold. I am a certified MBB in Six Sigma from Deloitte. I have been working primarily in Lean for about 26 years as an internal leader, a senior management consultant, and as an owner of three consulting companies over the past 15 years.

Bob Dodge: I completed several classes with BMGI in Lean [and] taught Intro to Lean to Vestas. I have facilitated Motorola and Northrop Grumman Change Management for Six Sigma BBs.

David Ringel: I learned Lean and Six Sigma in the automotive supply chain, serving Ford, GM, Chrysler [and] Mazda as a supplier of third-party packaging. I practiced Lean and Six Sigma (i.e., all CPI disciplines) as a process owner and change agent in the private industry. I also practiced CPI for 10 years as a performance improvement consultant, and I also hold the Lean Master and LSS/MBB certifications.

Joe Beggiani: I have used Lean Six Sigma to reduce process variation, decrease cost, and maximize efficiency. I was Lean Six Sigma certified in 2006.

Lynn Doupsas: I have completed some Prosci-related coursework. I have worked with Six Sigma consultants, but I am not certified as a GB/BB.

Scott Leek: I'm a practitioner of the improvement arts and sciences for more than 30 years, the last 17 devoted to Lean Six Sigma.

Sarah Carleton: I am a certified MBB. I have trained over a thousand GBs, BBs, and MBBs, over a span of about 12 years.

Roxanne Britland: I started in the world of Lean Six Sigma in 1993 with the original deployment at Allied Signal. My discipline in the area was about deployment strategy, and I built the product that became what's now called Initialization. That product took care of infrastructure components of a Lean Six Sigma deployment, such as human resource practices, the information management tools, the finance tools, project tracking, and communications components.

Darryl Bonadio: I have a DMAIC BB, DFSS BB, and an MBB. All were achieved through Juran, International. My DMAIC and DFSS BBs were in 2005 and 2008, respectively, and my MBB was in 2010.

Don Linsenmann: I led the Six Sigma transformation at DuPont for 16 years. I'm a certified GB and champion.

Bob Von Der Linn: I am GE Six Sigma BB trained, taught Six Sigma GB level for GE Capital, and have coached dozens of projects. I've had Lean training at GE and other companies. It is important to note that virtually all of my hands-on experience is in financial services—not manufacturing. Way before we started calling it Lean, in financial services, we were doing Action Workouts (i.e., process mapping, identifying value/nonvalue steps, and rapid implementation of pilot improvements).

Marc Zaban Fogel: I am GB trained. I've run champion sessions and/or Lean Six Sigma overviews in the past. My focus is mostly on the project selection and definition phase of projects, bringing in MBBs to do training and coaching in LSS.

Beth Cudney: I have written six books on Lean Six Sigma. I teach Lean and Six Sigma at a university and I regularly provide Lean Six Sigma training. I also hold the following certifications:
- IIE certified MBB, 2013
- ASEM certified Professional Engineering Manager, 2011

- ASQ certified Quality Technician, 2008
- ASQ certified Quality Process Analyst, 2008
- ASQ certified Quality Inspector, 2008
- ASQ Manager of Quality/Operational Excellence, 2007
- ASQ certified Quality Improvement Associate, 2007
- ASQ certified Six Sigma GB, 2007
- ASQ certified Six Sigma BB, 2005
- ASQ certified Quality Engineer, 2003

Currin Cooper: I attended Lean Six Sigma training while at Bank of America. I was already GB certified, requiring being a PM on a project that created documented revenues or savings of at least $250,000.

Mike Ensby: I have been an MBB-level trainer since 1998, and I completed GE's TTT course and spent 2 years training GE GBs, sales GBs, BBs, and MBBs. I also completed the Alcoa Business System (ABS) facilitator training, based on the Toyota Production System (TPS), which covers Lean/5S principles. Over the years, I have taught university-level courses and professional seminars using Lean Six Sigma content and have also been invited to speak on the topic at national conferences.

Patra Madden: I am an LSS GB and BB (Lockheed Martin certified), currently working on [an] MBB.

Scott McAllister: For almost 15 years I was in the Lean Six Sigma consulting space working at BMGI and had the privilege to work with more than 70 clients in 28 different countries.

Abdiel Salas: I hold both BB and MBB certifications (SBTI, Inc.), am a certified Lean manager (Ohio State University and Productivity, Inc.), and have developed a customized LSS for Invensys Corporation, where I train over 400 GBs and over 150 BBs in Mexico, the United States, and Europe.

Rick Rothermel: My experience with Lean Six Sigma is largely in assisting organizations with the integration of the Managed Change Methodology into their Lean Six Sigma approach.

Michele Quinn: I am a Lean Six Sigma MBB. I have led the deployment of Lean Six Sigma across several large global companies and have coached hundreds of belts, champions, and team members over the last 16 years.

Siobhan Pandya: I am a certified Lean Six Sigma BB. I have spent over 8 years in continuous improvement teams, including my current role as Director for CI/Lean at Mary Kay. In addition, I was a certified

Lean expert and CI business coach in Shell. I have deployed Lean Six Sigma throughout my career to deliver various benefits for Shell. I believe that the methodology used is not as important as ensuring that the tools are fit for purpose, so one day, I may be more focused on the Lean toolkit, and the next day, it may be the Six Sigma toolkit. This can also mean looking at alternative toolkits where necessary, e.g., TPM.

Q3. How would you explain the importance of change management to a Lean Six Sigma BB?

Tom Cluley: The technical elements of continuous improvement are the simple side of the equation. Sustainable change relies on creating the proper environment for success and establishing acceptance on the part of both the leadership team and the workforce. To establish that acceptance, you need to understand diversity and what motivates individuals and then learn to communicate in a manner that addresses their values and needs, not yours.

Bob Dodge: The return on investment of a great solution won't be realized if people don't use it. Change management increases acceptance, embracement, and sustainment of great solutions. Your team will produce ways to improve quality, but unless you address resistance, the people will not implement your solution.

David Ringel: There is a direct correlation between:
- the level of effectiveness of how a change is managed, start to finish
- the level of effort required to affect the change
- the level of benefit resulting from the change and
- the sustainability of the implemented change

Good change management can get more out of an imperfect solution than a perfect solution can get out of poor change management.

Joe Beggiani: Change management certifies that employees, customers, and stakeholders understand and support the change before it happens. Outcomes can be a reduction in the time needed to implement a change and improved productivity.

Scott Leek: Simple. Do you want your project to succeed or fail? Ignore change management at your peril. The road down continuous improvement is littered with the skeletons of improvement efforts of striking technical elegance [that] everyone in the organization dutifully ignored.

For the BB, change management is really about the WIIFM (What's in it for me?). And what's in it for the BB is project success. The BB should think this way and ask, "What's in it for people to make process improvement changes?" Change is not the natural state of a large organization.

So the BB needs to understand and, in some cases, experience firsthand that a perfect solution with little to no support or "acceptance" is doomed to languish and die of neglect.

Gary Bradt: I differentiate between change management and change leadership. Change management has more to do with a structured process, steps to go through...making sure that we're getting the focus groups together, making sure that we have a communication system in place, making sure that we have a feedback system in place. I put all those types of things—processes and procedures and activities—into the bucket of change management. Very important!

The other part is change leadership. Change management outlines the steps you go through. Change leadership is how you get the hearts and minds of the people that you're trying to reach. So my answer to your question is: understand the difference between change management and change leadership. Recognize when you're working on one versus the other and that they are both very important.

Sarah Carleton: Change management is an integral aspect of Lean Six Sigma as it appears especially in the WE space of the AQAL model of integral theory. If anything is to be accomplished, change must be managed. In almost every Lean Six Sigma project, significant change occurs, and BBs must be prepared to recognize, identify, understand, respect, and deal with aspects of change that affect their projects.

Roxanne Britland: I would look at it in two ways. First, there's a change management component about the introduction of Lean Six Sigma into a business. It is a new, structured deployment method that yields very strong results, project by project. And, just the idea of deploying Lean Six Sigma is a change for the organization. Second, project by project, improvement by improvement, you will also be introducing change to people's jobs, so depending on the person's role as a BB, MBB, trainer, or project lead, he/she needs to recognize that change is being introduced—possibly in two unique ways.

Darryl Bonadio: Unless you are able to have the improvements stick and become the new process, you are only performing part of your role. With process improvement, you can get implementation, but with change management, you get institutionalization. Institutionalization is when the process changes are embraced and owned by the people and become their new way of doing things. With implementation, you have only forced them into a new way that will not be sustained unless the changes are supported through reinforcement to drive preferred behaviors.

Don Linsenmann: With all of the statistics, processes, methodologies, and technology that you learned as a Lean Six Sigma BB, the lasting impact of your work will be dependent on how well your organization adapts to change. Change management skills will let you interact with all levels at your company, to communicate and solidify the changes you design into your control plan.

Bob Von Der Linn: There are numerous examples throughout history of people resisting new ways of doing things, even when faced with compelling logic and evidence that the change was beneficial. If you don't understand the human side of change, there is significant risk of failure.

Marc Zaban Fogel: Lack of change management thinking is THE reason good solutions at ANY scale fail.

Beth Cudney: A Lean Six Sigma BB must understand change management for the initiative to be successful. While the statistical analysis required to conduct a project is important, results will not be achieved or sustained without understanding and changing the culture of the organization. A BB must anticipate the roadblocks and fears associated with implementing a new strategy, so as to alleviate these fears.

Currin Cooper:
- Six Sigma is about consistency, error reduction and elimination, efficiency, and the ability to measure what you are changing.
- Project management is about delivering a project and keeping the project resources aligned to the schedule, budget, and scope.
- Change management is about business readiness—identifying and addressing the impacts of implementing the new strategy, initiative, or Lean process on the organization.

Note: Knowing about change is not the same thing as being prepared for change. Readiness isn't a newsletter or a survey. Change management is not the same thing as communications.

Mike Ensby: All the use of the LSS methodology and tools is generally useless in business process problem solving without an embedded change management approach baked into the structure. A sufficient stakeholder analysis at the outset of the problem definition should include a connection to scope elements, with an initial risk assessment that has a modicum of "barriers and resistance to change."

Patra Madden: I firmly believe we MUST incorporate CM tools and techniques into ANY LSS project from the very beginning. Many changes result from LSS projects, without understanding the stakeholders and their potential reactions. The resistance that surfaces while a process is being developed allows the team to deal with it and mitigate the damage. If you don't have the CM involvement, along with the voice of the stakeholder, the project is destined to fail before it gets going.

Scott McAllister: In the Lean Six Sigma movement, we focused an enormous volume of rigor on the quality of the solution, and I would explain that the right answer is not enough. Individuals have to change how they do their work for process improvement to succeed, and change management addresses the change that needs to happen at the individual level for any process improvement to sustain.

Abdiel Salas: As an LSS BB, you have to master change, and there is a psychology for changing people's behavior from state A to state B. It is a complex process. Also, you need to know that any change needs a stimulus. Stimulus is so important, because it determines the speed of change. Be careful not to manipulate, but to obtain, the "buy-in" change from people.

Rick Rothermel: Change management provides the focus on the people and cultural dimension of the changes and implementations that the Lean Six Sigma BB is responsible to deliver. Although Lean Six Sigma methodology recognizes the impact on the people, it is insufficient. The issues of individual and group resistance are typically ignored or underestimated, and they typically result in the installation of processes and/or process changes that are not fully understood, accepted, or adopted by the people who need to make the changes work, so as to deliver the business results the organization is expecting.

Michele Quinn: Lean Six Sigma is all about "Change for the Good," whether using the DMAIC process, DFSS process, or a Kaizen process. While having a new solution for a problem is a great outcome, without the adoption and usage of this new solution by the impacted groups and individuals, full realization of the benefits and ROI will not be achieved. This is where change management becomes critical.

Project Management can deliver the new solution, but change management is required to deliver the people side of change—adoption and usage of the new solution.

Jessica Bronzert: Change management brings a deep understanding of the human and organizational dynamics of change to an initiative. LSS, to my knowledge, does not have this focus to the same depth. In many cases, for LSS projects to be successful, the human side must be considered.

Siobhan Pandya: You can have change management without continuous improvement; however, you cannot have continuous improvement without change management. It is a requirement at every stage of a Lean Six Sigma project, and without the acknowledgment of its importance and the deployment throughout the entire project you will not be successful.

Q4. What is change management, in your own words?

Tom Cluley: Change management recognizes that in any population, there are a small number of individuals who are early adapters and an equally small number who are active resisters, with the bulk of individuals on the fence. To create organizational acceptance of change, you have to manage the transformation process, establishing WIIFMs. I've used different approaches in the past, but have settled on Kotter's eight steps, identified in his book, *Leading Change*, and building those steps into a Hoshin Process.

Bob Dodge: Change management is the process of identifying and reducing the root causes of resistance to great solutions.

David Ringel: To favorably reconcile all key stakeholder imperatives with respect to a circumstance that has been identified as "requiring change."

Joe Beggiani: Change management is a methodology and process that drives how we prepare, communicate, and assist employees to successfully adopt change.

Lynn Doupsas: I believe [that] ultimately, as a change manager, I deal with the people side of any change (system, culture, organization realignment, etc.). Change management is the use of techniques to help move people and organizations through the change curve.

Gary Bradt: Change management is putting together the systems, processes, and procedures we need to make sure an organization goes through a change process in a systematic and thoughtful manner, in a way that keeps everybody engaged and informed about what's going on.

Sarah Carleton: Change management is recognizing, identifying, understanding, respecting, and dealing with both the visible, outer aspects of change, and the invisible, inner aspects of transformation.

Roxanne Britland: It's the ability for an organization to introduce an improvement or a change that is accepted by all the players who have influence or are impacted by that change. So, it's mitigating the resistance through acceptance.

Darryl Bonadio: It is the institutionalization of action modifications within the organization that is evidenced through the observation of desired behaviors necessary to sustain the changes so as to achieve the goals of the organization.

Don Linsenmann: Enabling an organization to accept the different actions, results, and culture resulting from transformational efforts to improve the enterprise.

Bob Von Der Linn: The systematic identification and remediation of people-related risks.

Marc Zaban Fogel: Understanding the impact of your change on stakeholders and managing their needs and challenges as an integral part of your solution.

Beth Cudney: Change management is a strategy of understanding individuals' fears or resistance to change and addressing these fears to ensure that changes are smoothly implemented for the good of the organization.

Currin Cooper: You can't change what you don't understand. Change management helps to close the gaps between the "as is" and the "to be." Important projects, like BB or Lean efforts, can be disruptive. By their nature, they change the way work gets done. When you change the way work gets done, this requires new standards around process and procedures, terminology, roles and responsibilities, and very often, a change of culture. I believe that change management is more than "the people side of change." It is the secret sauce for mitigating risks when so much is at stake.

Basically, it boils down to:
- Do they know about the change?
- Do they know what to do?
- Do they know when to do it?
- Are they prepared?

Finally, will they do it? Who wants it? And who can get in the way?

Projects can be derailed when there's a mutiny or when a key constituency puts up a road block that leadership or the program office didn't see coming, but that was common knowledge in the rank and file. Telling people about what's changing is not the same thing as preparing them for change.

Mike Ensby: People clamor for change, but they don't want to be changed. Change management is the component of structured problem solving that clearly identifies "change-based risks," with the same level of rigor as is given [to] the technical aspect of the project. Specifically, a good change management approach is incorporated in the voice of the customer (VOC)-to-critical to quality (CTQ)-to-Scope.

Patra Madden: Change management is thinking about the people first and the project (or outcome) second.

Scott McAllister: It is a discipline and a management philosophy centered around helping individuals and organizations adopt new ideas. I have a strong bias that you only use change management to improve the outcome. It's not a "let's make everyone feel better" kind of approach, so while I think the discipline gets a bad rap for being touchy-feely, investing in change management should have a direct and positive output in terms of the effectiveness of the change you're working.

Abdiel Salas: Change management is a process to get to the next level of excellence.

Rick Rothermel: Change management is a focus on the people who are impacted by and/or can influence successful implementation, acceptance, and adoption of changes happening within an organization. Executives, managers, and supervisors who are responsible for leading change through their organization must leverage a disciplined methodology provided by change management to understand and shift the beliefs and behaviors of the individuals impacted by the change and address the resistance created as a result of change for the organization to successfully realize the intended results driving their decisions.

Michele Quinn: Change management is a structured process for supporting individuals and groups from the current state to a new, future state to deliver organizational results.

Jessica Bronzert: Change management is the discipline devoted to architecting understanding, alignment, and commitment to changes. This is about more than communication and training. It's about being clear about what change looks like when it has delivered on its promise, engaging employee mindsets and behaviors and continuing to stay engaged with the work until that original promise has been fulfilled.

Siobhan Pandya: The "what" we have to do is usually well defined; however, the "how we get there" is either undefined or unclear. Therefore, for me, change management provides the pathway to [being] able to achieve what we need to do, while taking into consideration every aspect of the journey, e.g., process, people, politics, etc.

Q5. What three words would you use to describe change management?
(Note that some respondents needed more than three words in their response.)

Tom Cluley: Gaining peoples' acceptance

Bob Dodge: Resistance mitigation process

David Ringel: Influence without authority

Joe Beggiani: Planning, implementing, communicating

Lynn Doupsas: Challenging, transformational, rewarding (when achieved)

Scott Leek: Purpose, people, process

Gary Bradt: Important, structured, thoughtful

Sarah Carleton: Recognize, respect, reconcile

Roxanne Britland: Results = Quality Solution × Acceptance (four words)

Darryl Bonadio: Behaviors, institutionalization, sustainment

Don Linsenmann: Transformation, acceptance, results

Bob Von Der Linn: Essential, challenging, leadership

Marc Zaban Fogel: Contextual, relational, ongoing

Beth Cudney: Culture, strategy, communication

Currin Cooper: Plan, identify, mitigate

Mike Ensby: Prior proper preparation

Patra Madden: Invaluable, intuitive, inspirational

Scott McAllister: Individuals adopting new things (with "new things" being one word)

Abdiel Salas: Challenge to improve, challenge to try, courage to defend change

Rick Rothermel: People risk mitigation

Michele Quinn: Change Management = Adoption and Usage

Jessica Bronzert: Nascent, underutilized, misunderstood

Siobhan Pandya: Critical success factor, sales, sustainability

Q6. Why use change management?

Tom Cluley: The technical aspects of continuous improvement are easy. Creating acceptance of change is hard and has to be managed.

Bob Dodge: Good change management provides a way for organizations to implement changes faster, reach better implementation results, and sustain the changes with far less stress than without it.

David Ringel: To increase the return on the investment of resources required to affect a change, as well as to increase the sustainability of the change.

Joe Beggiani: Change management can help an organization achieve the people side of that change. A new process, a new system, or even a new organization design is no good if the people who are going to be using it are not bought in. Without change management, change resistance, lower productivity levels, and employee turnover can, and most likely will, take place.

Lynn Doupsas: Most large changes fail if change management techniques are not used.

Gary Bradt: Because if you don't use change management, you'll have chaos.

Sarah Carleton: As an essential aspect of any project, change management increases the probability of success of any project and helps everyone involved with the project grow.

Roxanne Britland: As we evolve in our careers and become managers or leaders in any capacity, I think there needs to be an awareness that there is a method and technique and principles to be taught, learned, and practiced.

Darryl Bonadio: Because without "winning the hearts and minds" of the people, you will not be able to sustain the changes, and things will deteriorate rapidly after oversight activities cease. All organizational changes are really about getting people to accept and own the new activities with a positive attitude that doing things this way will result in benefits. Those benefits span individual and corporate values and, as such, need to be communicated in multiple levels of meaning to the employees. Change management helps you identify how the changes will impact people, the behaviors needed to sustain the changes, and how to assure the right reinforcements are implemented to assure change.

Don Linsenmann: It is the complementary "technology" that supports the technology of LSS.

Bob Von Der Linn: Simple; to mitigate risk and maximize the ROI of an initiative.

Beth Cudney: To ensure a smooth transition.

Currin Cooper: Change management can effectively target and mobilize groups and individuals and form organizational partnerships. This allows key stakeholders or leadership to focus limited resources on the right activities, goals, and outcomes.

Mike Ensby: The alternative is to achieve less of a result in addressing the business process issue at hand. Effective change management helps reduce the "sag" principle of traditional PDCA, LSS, and KBM methodologies. The most variable variable in organizational problem solving is people. Change management helps keep the cats herded.

Patra Madden: To improve the life of your coworkers, your customers, and yourself.

Scott McAllister: To improve results and outcomes and sustainability of our solutions.

Abdiel Salas: We are part of a whole organization; therefore, our behavior affects the whole organization's results. Because people are important, we need to understand change beyond our own world and get a bigger picture. By using change management, we want to diminish harm and pain but still speed up and assure the results.

Rick Rothermel: Use change management to proactively predict, identify, validate, and mitigate the resistance expressed by those individuals impacted by change, enabling the organization to achieve the intended business results that generated the need for change.

Michele Quinn: Without adoption and usage of the "new way" by individuals and groups, organizations risk limited benefits achieved, a growing culture of failed change, and frustration impacting customers and employees.

Jessica Bronzert: The vast majority of large-scale changes attempted in organizations fail to take into consideration any meaningful understanding of the human and organizational dynamics of change and therefore fail to deliver sufficient results. Change management, when applied by seasoned practitioners, brings this critical lens to the table to increase the chances of success.

Siobhan Pandya: On the basis of the three words mentioned earlier (critical success factor, sales, sustainability), you will not be successful without change management, as it will highlight areas for attention that project management alone will not uncover, e.g., the level of support of staff, the impact that staff has on the outcome of the project, etc. Also, it will help to define the sales pitch that you need as a project lead to help everyone impacted by a project understand the importance to the organization, but more so, why it is important for individuals, what it means for them, and how they contribute to the bigger picture. Change management helps you focus on each individual, thus ensuring that you have support and long-term sustainability.

Q7. How have you seen change management used in conjunction with Lean Six Sigma? Was it successful?

Tom Cluley: My companies have used change management in hundreds of client engagements at both the program and project level. Success varied, depending on leadership engagement.

Bob Dodge: Motorola and Caterpillar are great examples of the benefits of the combination. Their application was at both levels.

David Ringel: Yes, Yes, Both.

Joe Beggiani: Yes, I have used change management on a project on which I was the leader. I had a Six Sigma leader and a change team. The project was successful, because we brought the change team in at the very beginning. We had a very strong sponsor, which was an enormous help in the communication portion of the change. But also, I was able to focus on the day-to-day project build and implementation, while the change manager focused on the training and communications delivery plans. Our Six Sigma team focused on the processes to be used, and this integration of a focused team eliminated a number of risks, brought some great ideas to the table, and helped us use the end user as part of the testing team. This helped with adoption, employee readiness, and feedback.

Scott Leek: Yes. Yes. Program and project level. Both have a role in my opinion.

Sarah Carleton: At both the program and project levels, change management is used as an essential aspect of LSS. It is most successful when change management has been taught to BBs, and it is expected that they use it and report on it in their project reviews.

Roxanne Britland: Let me talk about it first at the project level. It was my experience for over 15 years that many of the MBBs who were developing curriculum about training belts would bring in a module called change management. It might be a 2- to 3-h block, and it was a one-way knowledge transfer delivery. I'm not saying some didn't do more, but mostly it was just checked off the list as a training module. Change management is not something that's merely trained; it needs to be applied. It needs to be practiced to the benefit of the organization introducing the change.

From a leadership perspective, those companies that I worked with were endorsing Six Sigma from the top and investing a lot of money in the idea that Six Sigma was going to be their improvement methodology. My personal skills in helping them deploy Six Sigma was really around making sure that the infrastructure was well established so that you addressed the people needs, the finance needs, the IT needs, and the communication needs. In those organizations that did *not* deploy from the top down, change management was absolutely required, but in many cases, was not addressed.

Darryl Bonadio: I have been personally involved, either directly or as a coach, in over 100 different Six Sigma projects and have seen how change management has impacted the success. Depending on the size of the efforts, we have used change management at both the project and program level. Key to change management is the support and reinforcement of the changes by management. Successes have been great, but failures are as spectacular. If management does not understand and embrace the changes and know how these changes will impact the people, they will assume everyone will simply accept the new. Without positive and negative reinforcement of the changes, employees will receive mixed signals and assume that the changes were either not supported, or worse, that no one cared what they thought.

Don Linsenmann: Yes. Yes. Used at the program level and then cascaded to the project level.

Bob Von Der Linn: In the late 1990s at GE, CAP was an integral part of GB and BB training. You could not do a Six Sigma project without incorporating CAP. Over time, it became obvious that change had to be managed at a program level.

In 2016, change had to be managed at an enterprise level. What Jack Welch understood in 1990 when he commissioned the development of CAP has never been truer (he was a genius). In the typical enterprise today, managing "changes" is a fool's errand. You must enable your organization to manage change. It is constant, and there are interdependencies between initiatives that make it impossible to address them independently.

Marc Zaban Fogel: Most programs I've been a part of (Six Sigma and otherwise) fail to apply any change management beyond basic project definition and alignment of the team, and so [they] struggle at the implementation phase of the solution(s). Where it is applied well, it's seen as an indirect support to the project or program.

Beth Cudney: Change management was provided to us as additional training during the BB course. Since we were given the task of leading the change throughout the organizations, we were fortunate to be provided with additional training solely on change management. It was successful in changing the culture of the organization. It was at the program level.

Mike Ensby: If it is begun at the outset of the program/project, with a clear connection to the "stakeholder-scope-risk" discussion, the champion can be better informed on the "temperature of the room" and better lead the communication offensive that should be part of every phase of whatever methodology/model is being leveraged. The sooner the change strategy becomes a centerpiece of the DMAIC/DFSS/5S process, the better the probability of a successful implementation of the ultimate solution, and the longer the gain(s) can be held. It's the difference between getting commitment over participation from the problem-solving team and the various stakeholders. Accountability for the "change" cannot be delegated.

Patra Madden: Everything I do utilizes change management. When running a Lean Six Sigma event or project, I ensure the identification of stakeholders is robust. During the initiative, I often ask the team, "What do you think the _____ will think of this?" The impact on the stakeholder of the changes identified in the Lean event is discussed before, during, and after the mapping of the new process. I believe one of the keys to a successful process change is identifying the impact on the stakeholders and determining resistance mitigation methods. It doesn't matter if the team determines the impact has a negative or positive effect. What matters is the perception the stakeholders have and that the team acknowledges it.

Scott McAllister: The short answer is "yes." And, I've seen it where the answer is "no" as well. We spent almost 13 years using basically the same approach, in terms of structure, methodology, and tool set, and some clients were wildly successful, others were modestly successful, and some didn't achieve any success at all. It wasn't because of technical aspects… those stayed the same. It was all tied to the adoption of the approach. It's not just the tools. It's often-times the mindset, so change management, both at the project and the program level, is an absolute necessity and to me is what separates the winners and the losers.

Abdiel Salas: Any LSS project or program should be executed before we see the result; therefore, all LSS projects and programs need change management to be successful.

Michele Quinn: Absolutely. Starting in the initial project definition and chartering, effective change management should be integrated at each step/stage of the process.

Siobhan Pandya: See my reference above that LSS cannot be successful without understanding and deploying change management. Change management helps you recognize how you manage your different stakeholders, helps you understand their attitude toward the change, how they can influence others, what is important to them, which ones to focus on and to what level, and how to ensure they help you succeed in your project. When done correctly, you will be successful in your project—I have used it at the program and project level.

Q8. What do you think the typical Lean Six Sigma BB needs to know about change management? How should it be taught to the belts?

Tom Cluley: Typical Lean Six Sigma BBs should become experts in change management, if they wish for sustainability regarding their improvements. It is best taught in a train-and-mentor approach, with training followed up by mentoring in which the tools are called on situationally. I find it follows a learning curve of (1) unconscious incompetence, (2) conscious incompetence, (3) conscious competence, and (4) unconscious competence. This transitioning only comes with experience.

Bob Dodge: They need to know it will make them far more successful, valuable, and fulfilled. It should be taught only if the practitioners and their sponsors are committed to use it. Otherwise, time and resources are wasted.

David Ringel: Relationships trump reason, emotions trump logic, culture trumps individuals, and data trumps conjecture. Personal styles affect how messaging is both given and received, whether there is any perceived value in the message—learn about personal style and how to apply situational behavior/communication as a facilitator/change agent. Learn to listen effectively and seek to understand the rest of the story. Make no assumptions, develop no opinions, and alienate no stakeholder. Learn to effectively facilitate consensus between those who otherwise want it their way.

Joe Beggiani: Based on my background, and being a LaMarsh Change Master, I would say they need to understand the change model ... identify, prepare, plan, implement, and monitor. Others' methodologies will say prepare, manage, reinforce. What is entailed in these sections, impact assessments, sponsor screening, communication plans, training, etc. They need to understand this runs in parallel with the project management and/or DMAIC life cycle and are enablers to success.

Scott Leek:
1. Understand the basic concepts, including the need for change management
2. How to use the basic tools to "grease the skids" for project success
3. Some basic classroom training
4. Learning mostly on the job with a mentor or coach

Sarah Carleton: BBs must be prepared to recognize, identify, understand, respect, and deal with aspects of change that affect their projects. CM should be taught as an integral aspect of every phase of the Lean Six Sigma methodology.

Roxanne Britland: I think they should be certified in a change management methodology. There are programs available. I don't believe CM is a PowerPoint set of slides that has been handed down, company by company, MBB by MBB. It is a religiously practiced method and approach, project by project. There needs to be a change management plan. It's not something you get from PowerPoint slides. It is a trained, and then an applied and practiced, method of introduction. I also don't believe that it's a stand-alone activity. I'm not too sure I really understand organizations that say, "I'm going to go buy change management." All of a sudden, they're training a lot of people and sending people to seminars. There really have to be applications and related activities. There has to be a real change, a real solution that's being developed using an applied method for managing change.

Darryl Bonadio: We use what we call mindsets and behaviors. What we teach is that there is a human side to all changes, and if you do not address the needs and feelings of the employees, you will not get your improvement to be sustained.

Don Linsenmann: Influence skills... Communication skills... Stakeholder analysis... Process model. Taught at BB training.

Bob Von Der Linn: Every BB should have training in organizational behavior and change management, and then should be given hands-on experience facilitating change management tools.

Every BB should understand that to sustain a change (control), you must realign the business systems and structures with the desired end-state. If you do not update your measurement and reward systems, hiring and staffing systems, training and development systems, etc., those systems will push you back to the old way—that's what they are designed to do.

Marc Zaban Fogel: All belts should be taught the need to incorporate stakeholder needs into their thinking and planning. It can be most simply done as stakeholder analysis and communications planning but needs to be emphasized as an analysis tool to target barriers before they come up. It should be brought up at the beginning of a project and discussed actively when the FMEA concept is taught as a tool.

Beth Cudney: How to effectively communicate change in terms of what is important to each individual. They must understand individuals' potential fears and what's in it for each individual. It should be taught consecutively with each tool in the Lean Six Sigma program.

Currin Cooper: Implementing a new process without understanding the impacts or unintended consequences is risky. Very few lines of business or business units actually do work in a silo. Organizations are systems and processes are cross-functional. Politics are real. You can't influence what you don't understand, and you can't understand that of which you aren't aware.

Mike Ensby: Much more than is typically covered. There is more emphasis on the cadence of the application of the "tools of problem solving" which are inherently disruptive to existing processes, than there is on measuring the impact of the change to the status quo. Most belts are not people-change facilitators. They need a basic, "freeze–unfreeze–refreeze" model to work from.

Patra Madden: I believe all BBs should be certified as CMPs prior to completing their LSS training and BB certification. I also believe all BBs should be trained in cause mapping (root cause analysis), conflict management, effective listening, etc.

Scott McAllister: I think it needs to be integrated into Lean Six Sigma training, so it's not different than, but is a key part of, the program. And, I think they need to learn a model for organizational change and for individual change and understand when they use which parts of the approach.

Abdiel Salas: Be prepared to work with all levels of resistance, teach the basics of [the] change management process. Increase the change management topics.

Rick Rothermel: Lean Six Sigma BBs need to understand the value of the acceptance and adoption of the change by those individuals who are impacted. They need to understand what factors generate resistance and the strategies and tactics they can use to proactively mitigate the resistance. Learning change management methodology and tool application should be integrated into the Lean Six Sigma BB curriculum. It should align to the people requirements and be delivered as a part of the project implementation.

Michele Quinn: I'd start with the ADKAR model as a foundational model for individual change. Building from here, BBs and others need a solid process and set of tools to support their ability to deliver both the new optimized process and the people adoption and usage of the new optimized process.

Jessica Bronzert: Process professionals generally can benefit from understanding the human and organizational dynamics of change.

Siobhan Pandya: The main thing that they need to know is their stakeholders' attitude to change—from there they can determine how best to approach them and win their support. They should learn the basics about the various models—Rogers Curve, Bridges Model, etc.—through theory, but also simulation, e.g., role play with folks in different stages of change, and learn how to react and navigate conversation.

Q9. What are the critical success factors, i.e., the secrets to success, in using change management in conjunction with Lean Six Sigma?

Tom Cluley: The critical success factors are measured in terms of:
1. Organizational urgency
2. Will to change
3. Awareness of deficiencies
4. Skills required
5. Reliability

These are enforced through the proper:
1. Leadership
2. Empowerment
3. Communication
4. Learning
5. Discipline

Bob Dodge: First and foremost, the sponsors of the initiatives, programs, and projects must practice good change leadership. One must understand that it is an ongoing practice and not just a "punch list."

David Ringel: Other than [my] previous response [to Question 8], the ratio of the level of effort put against change management and the technical solution should be around 4:1.

Joe Beggiani:
- Committed sponsors
- Change leadership team
- Buy-in from the senior and frontline management
- Continuous and targeted communications
- Sustaining the change after project completion

Lynn Doupsas:
- Leadership alignment
- Clear understanding of the end goal
- Employ tactics that lead to increased adoption

Scott Leek:
- The tired cliché of senior leadership: INVOLVEMENT
- A worthy project (worthy = solid business case + people care + customer impact)
- Project leader with the competencies to lead a project and inspire change
- Sincere commitment to addressing people's legitimate concerns

- An organizational culture committed to not only getting it done, but also, doing it better

Gary Bradt: (1) That you have the buy-in and sponsorship of senior leadership that has the ability to get things done. (2) That you have a communication system that is ongoing and not just a one-time announcement, as well as a way to gather information about how people are experiencing the change—two-way communication, not just top down. We also need bottom-up communication. (3) We need to capture the hearts and the minds of the people who are being impacted by the change. (4) We need to recognize that the people who are leading the project on the change are going to be living and breathing that project, way more than anybody else who is being impacted by it, and we need to keep in mind, if you're leading the change management project, you have to give people time to catch up with you and the time to think through all of the issues that they might not like about it. You must take the time to address those issues. Many times, people who are leading change management efforts have lived, breathed, and eaten that concept for months, and when they're ready to roll it out, they're shocked that people don't immediately buy in.

Sarah Carleton: BBs should ask these questions at every phase:
- What is changing in this phase?
- What does this mean to me?
- What does this mean to stakeholders?
- What does this mean to the team and the organization's culture?
- What individual skills have to grow to accommodate this change?
- What organizational systems have to grow to accommodate this change?

Roxanne Britland: Those that are training Lean Six Sigma practitioners in change management need to be certified and practiced in it—and not just a transfer of information on slides. It is a practiced methodology. Leadership and deployment champions need to endorse change management as a serious element of their Lean Six Sigma deployment. It's not a "sprinkle in"… it's embedded. It's got to be a part of the way you do business. And Lean Six Sigma says that all the time, right? We're going to be just the new way of doing business when you want to improve something. Change management needs to be embedded with that same mindset.

Darryl Bonadio: Involve key players in the change. Understand the impact of the process changes on the individuals. Make sure that in addition to the control plan there are a set of reinforcing behaviors that need to be supported.

Don Linsenmann: Top management support… Recognition of a culture change is needed… Results measurement… Open communication.

Bob Von Der Linn:
1. Engage with a change consultant early.
2. Engage with the managers in the affected population early and often.

Marc Zaban Fogel:
1. Ongoing to analysis of stakeholders and communication adjustment by the project manager
2. Ongoing discussion by the champion of resistance to change
3. Willingness of the project manager and champion to engage or challenge key people who resist the change

Beth Cudney: Teach it as part of the training curriculum.

Mike Ensby: Change management is the "critical" aspect of LSS efforts. Projects are about changing the current state. So, success is predicated on a properly developed change approach.

Patra Madden: To not take every tool at 100% face value.

Scott McAllister: Embedding it early in the process. We've got a strong bias within Prosci. We're trying to get individuals to the Ability milestone in their ADKAR training, and it saves time when we're training on solutions. Successful adoption of the solution requires more than just communication and training.

Abdiel Salas: We need to know that all good ideas are only ideas if they are not implemented and that once they are implemented they become innovations. There is a difference between my idea and our idea, go and try. Resistance to change is normal behavior, but if there is no will to change, then you are destined to become obsolete.

Rick Rothermel: "It's not about the tools!" Like any disciplined approach, change management methodology and tools provide the framework for data collection. The critical success factor is in the analysis of the data and the recommendations implemented to increase the probability of success. Remember, there are two important aspects to success in the use of change management with Lean Six Sigma: Six Sigma supports the identification and implementation of the best solution/process.

Change management supports the acceptance and adoption of the solution/process by those who are impacted. You need both these factors working in unison to achieve the intended business results.

Michele Quinn: Understand the ADKAR model, be willing to focus as much effort on supporting individuals through awareness, desire, knowledge, ability, and reinforcement as the effort we put into value stream mapping and variation/waste analysis.

Siobhan Pandya: You must know what is important to each individual stakeholder—whether it is a frontline employee or the CEO; each individual is motivated differently and engaged differently, and therefore, understanding this will help you achieve the improvement opportunity you are seeking. It will also help you determine the best solutions as you will have support and buy in from the beginning. This will also help with sustainability.

Q10. Lessons learned in using change management (with or without Lean Six Sigma)?

Tom Cluley: Lessons learned are that change needs to be managed and that without engagement on the part of leaders, results will be marginalized.

Bob Dodge: Change sponsorship (leadership) is as critical as change management. In my experience, many leaders try to delegate both, and it won't work.

David Ringel: Don't leave home without it. Start with CM; don't stop; end with CM.

Joe Beggiani:
- Get change and program managers together early in [the] planning phase
- Communicate expectations often and to everyone
- Reward for the change
- Change is hard; understand the change and the impacted groups
- Answer the employees' questions
- Put the time in to sustain the change

Scott Leek:
1. If there is no or insufficient communication about possible or certain changes, stories are created to fill the void, and almost always paint a more dire picture than is reality.
2. Seek first to understand, but know that understanding sometimes comes from observing behavior, not listening to words.

3. Everyone's behavior makes sense to them.
4. Stay focused on behavior, not on trying to change attitudes.
5. There are legitimate fears and concerns, and then there is sabotage. Recognize the difference and act accordingly.

Sarah Carleton: Change management must be addressed as an integral aspect of Lean Six Sigma.

Roxanne Britland: Change management trainers or consultants really need to work closely with the business improvement people. It's not a separate track. Sometimes the change management folks are not in the project all the way through, so they don't really know what the solution is.

Darryl Bonadio: Management support, through its behaviors and attitudes, has a significant impact on success. When implementing across functional areas, you need to understand the key drivers within each functional area, so as to assure consistent support.

Bob Von Der Linn: As a change consultant, when the change is managed well, my client appears to be a highly effective leader and gets all the credit. When the change is not managed well, regardless of the root causes, the leader appears highly ineffective, and I get blamed.

Marc Zaban Fogel:
- If the champion or key stakeholders do not actively want change, you will fail.
- If you don't get and integrate feedback from frontline people, the change will fail.
- Taking a pilot/experimentation approach to implementing change(s) is the best mindset to finding effective solutions.

Beth Cudney: Communicate and get people involved early to help gain buy-in.

Currin Cooper: Be pragmatic. Don't go crazy with the forms and surveys.

Water cooler talk will happen. When information is lacking, people will fill in the gaps with misinformation. Change management helps get in front of that.

Mike Ensby: CM can never begin too early. Most LSS (or other) projects do not pay sufficient attention to the stakeholders—the analysis is too superficial. Just when you think you know how a stakeholder will respond, they will react in an entirely unpredictable manner. "That's not what I wanted/expected." Change is more difficult than most leaders understand. Just because a solution makes sense doesn't mean people will embrace the change.

Patra Madden: Just because a change appears good to one person, NEVER assume the rest of the people will like it. Everyone is different and deserves to be heard.

Scott McAllister: There's no such thing as perfect change management, so don't let perfect get in the way of progress. The more you invest on the people side, the better results and outcomes you get. We often call it the soft side of change, but it's actually the hardest part of making change happen.

Abdiel Salas: Most changes are not welcome because they are not understood. Once they are fully understood, they are mostly welcomed.

Rick Rothermel:
- Be clear on your direction—what are you trying to accomplish and what will the Future look like when you arrive?
- Leadership support, engagement, and ownership of the change. Sponsorship of the change is critical, and without it, the project will not achieve intended results.
- Have a good, solid plan and follow it.
- Recognize that there will always be a level of resistance to the change, no matter how great the change is. Be prepared for the resistance and be willing to take the necessary steps to mitigate the resistance.
- Align the change to the mission, vision, and goals of the organization. The change has to make sense to the organization and the employees.
- Recognize there are multiple changes happening in the organization at the same time, and your change may not be the "most important." You need to assist employees in calibrating the priority of a specific change.

Michele Quinn: Focus on the behaviors—those that are required in the current state, and those that will be required in the new future state. Apply the ADKAR model, identify barrier points, and use a proven structured CM process and methodology.

Jessica Bronzert: Transformational change is fundamentally messy, and there are known patterns of risk that contribute to whether the change will be successful or not. These include a lack of effective sponsorship of the change effort, an unclear picture of what the future looks like after the change is successful (or a lack of alignment among decision makers about the future), and a failure to continue supporting the change past installation until the full business benefits of the change are realized. Any change management approach that doesn't architect these outcomes from the beginning, and monitor and mitigate them as the change unfolds, will be less effective.

Additionally, the change management function itself must be well sponsored. If the leaders of the change itself are not bought in to using a change management approach and don't fully partner with change management practitioners supporting them, it will be very difficult, if not impossible, for change management to be an effective approach. Even when change management is fully supported, it can't be outsourced to practitioners alone. Sponsors and other stakeholders responsible for making the change successful must build their own capability in the change space as well.

Siobhan Pandya: Important to have [a] clear end goal defined, as you need to be able to communicate this to others. Often, the journey has been embarked on without this clarity. Take the time required to use change management—more work and more difficult at times, but better in the long run.

Q11. What can go wrong when using change management?

Tom Cluley: Failure to engage leadership in the process can lead to frustration, especially if the lower level workforce understands the principles and sees it lacking in their leaders.

Bob Dodge: Leaders trying to delegate change leadership/sponsorship. Disagreements over whether it is needed, being done effectively, who should do it, and how.

David Ringel: The greatest risk for a belt is to have, and to allow to come to light, an opinion or position on subject matter that is not theirs (theirs being the CPI tools, methods, agenda). Belts must be trusted as honest brokers of the change.

Joe Beggiani: There are a number of reasons—poor sponsorship, poor communication planning, not planning the change at the beginning of the project, cutting corners on the prepare phase of change. There are many more, but it does not have to be this way. If you have committed sponsors and a strong change team, you can make sure that a change plan will be strong, but only if the program/BB is open and committed to the change.

Scott Leek:
1. If handled ham-handedly, the effort can be perceived as nothing more than crass manipulation, instead of sincere attempts at dialogue to understand genuine concerns.
2. Loose credibility, if changes are not seen through to completion, or if changes are made upon changes, without proper assessment or evaluation.
3. Spending too much time planning for and overestimating the resistance to change. This is a particular risk if the change management professionals are not sufficiently grounded in operational realities and/or are otherwise underemployed.

Gary Bradt: (1) Under-communication. You have to have a simple message, and you need to communicate it repeatedly. (2) Not giving people a voice in the change process—taking it top down, [saying] just do it. (3) Not giving it enough time. Giving up too soon. (4) Declaring victory too soon.

Sarah Carleton: Change management can go wrong when insufficient attention is paid to it, when it is not taken seriously, and when it is used in an immature manner.

Roxanne Britland: Change for the sake of change. For example, a change agent who comes in and wants to introduce these methods and practices of change, but is not doing it in conjunction with a solution. Change management can't be a sidebar; it's got to be embedded. And, it shouldn't feel like you're doing it. It should just naturally be a part of the project.

Darryl Bonadio: A poor communications plan can cause more issues than solve. Knowing what to say, to whom, and when, can make a huge difference in how well the change is perceived and accepted.

Bob Von Der Linn: Almost universally, unenlightened leaders assume that a communication plan and a training plan are all that is necessary for change management.

The most significant problem, however, is the semi-enlightened leader who believes that he/she can "outsource" change management to HR or to a consultant (internal or external).

Beth Cudney: Not getting sufficient employee involvement and not communicating effectively. These will lead to resistance, the change will not occur, and the culture of the organization will be negatively affected. When this happens, executive management is often no longer trusted.

Currin Cooper: Spending too much time on non-value-added activities. Not helping sponsors see the value in the approach you are taking. (Not all practitioners understand it themselves, so it is difficult to help others to understand it!)

Mike Ensby: Trying to satisfy too many "wants" and "like to' s," and losing focus on the "musts." A lack of effective communication (under-and/or over-communicating). Treating all stakeholders equally. Not having a committed senior leader as the project champion. Everything and anything!

Patra Madden: Sometimes the team or sponsor wants to take shortcuts to complete a project and will often try to cut out the CM piece. Don't let them!

Scott McAllister: It's a great scapegoat when things don't work out to say we needed more change management, so understanding that it's a bit of a minefield to get started in. It takes some courage to take on a leadership role in this kind of work. A lot of people misunderstand communications and training as change management, so you often times are misunderstood, I think. Everyone has a different definition of what change management is, so creating a common definition and language around it is important. Oftentimes, it's a derailer when we think everyone is talking about the same thing, and we're actually talking about wildly different things.

Rick Rothermel: I believe any focus on the people impacted by the change is a good thing, even if it is minimal. When applying change management discipline, it is possible to place too great an emphasis on the tools and too little emphasis on the analysis that is required to determine solid and reasonable action steps. This adds time and complexity to the project, which may create more resistance than is mitigated. Keep it simple and reasonable. The intent is to understand the resistance and deal with it. Another pitfall is to misread the client or leadership. Many leaders don't want to address this, because they perceive it as "soft." Shift the conversation and the focus to risk. Change management is addressing the risk(s) associated with the people who are impacted by change.

Michele Quinn: Lack of effective leadership and sponsorship is the number one risk.

Jessica Bronzert: The worst thing that can happen is that the change itself fails to deliver its intended benefits. While the use of change management should contribute to the success of the change itself, ineffective or shallow change management efforts can, at best, be neutral in their impact or, at worst, contribute to the failure of the initiative to deliver. Strong change management goes well beyond the traditional thinking of "communication and training." There are known patterns that contribute to the success or failure of change initiatives. A change management approach that doesn't take these patterns into consideration is insufficient. Lack of sponsorship for the use of seasoned change practitioners using a robust methodology is insufficient. An approach that is only started "when implementation happens" is insufficient.

Siobhan Pandya: Taking too long with certain stakeholders, moving forward without clear support, [being] too caught up in emotions, etc.

Q12. What are the dos and don'ts with regard to change management?

Tom Cluley:

> Do—learn to recognize diversity in individuals and that what motivates one individual to accept change may not motivate others.
>
> Do—insist that while the workforce can be won over in time, at the onset, the leadership team must fully support the effort
>
> Do—require leaders (formal and informal) to lead the change. It can't be delegated.
>
> Do—use an assessment tool that measures the element of cultural change, so that you know where the gaps are and can focus on areas of need.
>
> Don't—assume that change will happen without being managed.
>
> Don't—accept resistance on the part of the management or leadership team.
>
> Don't—expect organizational change to come quickly. My experience is that it takes the best organizations 3 to 5 years to create organizational change. Transformation follows a very similar pattern to Collin's flywheel effect described in *Good to Great!*

Bob Dodge: In my world, in small businesses, I cannot try to "do" what worked for large enterprise clients. I have to think very carefully about where there is an opportunity to help these businesses make changes, based on the frameworks to which I have been exposed. The one piece that is reinforced over and over is to *Start with Why*.

Joe Beggiani: My major don't is assumptions… check all assumptions! Also, don't cut corners. Follow the methodology. It was put in place for a purpose. If you are the change leader or a sponsor of a change, be committed. You are driving the team making major decisions, and you need to be engaged. DO NOT force the change. Work through the change and get the end users involved if possible.

Gary Bradt:

Dos:
- Involve the people impacted by the change.
- Give them time to buy into the change.
- Communicate it until you are blue in the face.

Don'ts:
- Have a major meeting to announce the change and expect everybody to buy-in immediately.
- Make it just a top-down process.
- Get hung up on the people who are most resistant. I call this the 10-80-10 rule. About 10% of the people won't buy in, no matter what. About 10% want change, no matter what. It's the middle 80% that you need to focus on. If they are properly led, with proper change leadership and proper change management, you'll get them to move along. The bottom 10% are the most vocal and negative. Change leaders will get hung up trying to address the vocal minority and miss the majority of the middle just waiting to be led.

Sarah Carleton: Do recognize, identify, understand, respect, and deal with aspects of change at every phase of the project. Do involve team members and stakeholders in change management. Do review change management aspects at every phase review.

Don't gloss over any aspect of change management. Don't treat it as something separate from Lean Six Sigma.

Roxanne Britland: Do role model good quality results that are accepted by the people who are going to be effected by that change/improvement. Do recognize that change occurs every day. It's not something special. New people receiving training is a change. New skills are a change. Improvement is change.

Don't think you can communicate your way through effective change.

Darryl Bonadio: Dos: Over-communicate. Understand what drives behaviors within the target community. Sell the change, even if you think it should sell itself. Get people involved so they own the change.

Don'ts: Assume everyone understands why. Think that the existing reinforcements will support the change. Confuse change with good.

Marc Zaban Fogel:

Do:
- Make the entire project team aware of and responsible for change management
- Continually collect and review stakeholders feedback
- Find the people who are willing to try and make them successful first
- Expect resistance as a natural reaction

Don't:
- Stop with the initial change management plan
- Ignore frontline people with low power
- Expect logic and facts to explain all things

Beth Cudney:

Dos:
- Communicate, communicate, and then communicate.
- Get feedback before starting a project or an initiative. People want to be involved.
- Management must walk the walk and not just talk the talk.

Don'ts:
- Handle every employee the same way when it comes to change. Each person may have a different fear. They all must be addressed with an effective communication strategy.
- "Sell" the change. People need to feel it and understand the benefits of the change.

Currin Cooper:

Do be an advocate for the employees/stakeholders who will have to live with the "new way".

Don't fall into the trap that change management is sending out communications. It is consulting, coaching, asking difficult questions, driving out gaps, and anticipating challenges.

Patra Madden:

Don't assume you don't need it. You always do! (Not always at the same amount or the same thing. But always something.)

Scott McAllister:

Do—start early, and make sure you engage with sponsors early in the process, and have a budget for it

Don't—think you can send an email on Monday, with training on Tuesday, to go live on Wednesday, and be successful. Let perfect get in the way of progress.

Abdiel Salas:

Do—involve. Welcome all ideas. Accept criticism. Accept wrong. Expend time to explain. Lead change. Look for the better of most people. Appreciate by listening. Use inclusive words. Use personal names, not titles. Make jokes, but be firmed to ideas. Make team decisions. Care about the people more than processes. Be prepared to give back something to people, always quickly. Always thank and appreciate the help of others. Be kind and be responsible for failure.

Don't—impose ideas. Never promise more. Do not ignore or misspell names. Never kill ideas. Never ignore small ideas—see ideas as opportunities to be at the next level of acceptance. Never show you don't care. Never manipulate or have hidden agendas. Don't give the impression you are the superstar hero.

Rick Rothermel:

Do—collect data upon which to make people impact decisions and recommendations.

Do—engage leadership, and force them to accept their role and associated responsibilities for leading change.

Do—recognize that there will always be a degree of resistance by those impacted by the change, regardless of how great or positive the change is.

Do—recognize that change management is a disciplined approach, but unlike project management, it is not a linear application. It is simultaneous and iterative in the implementation of tasks and outcomes.

Do—make certain leaders understand they must provide positive reinforcement when they see change happening AND deliver corrective action when it doesn't happen.

Don't—make it about the tools. Tools are intended to support the data collection. The real value is in the data analysis and recommendations implemented.

Don't—allow leaders/sponsors to abdicate their responsibilities. They need to be seen as the leaders of the change.

Don't—make it more complex than is necessary. The level of change management support required is driven by the resistance to the change, not a templated project plan.

Michele Quinn: Coach and guide sponsors/champions, engage middle management early, and never underestimate how human beings will cling to the current state, even when unsatisfied. Provide a clear vision. Communicate well-articulated answers to the questions, "Why this? Why now? What's the risk of not changing? What is expected of me? What's in it for me?"

Siobhan Pandya:
Do have a clear plan, clear communication tailored for different audiences, stakeholder management, empowerment.

Q13. Do you have a favorite change management methodology? If yes, please explain.

Tom Cluley:
Employing Kotter's eight steps into whatever process I am trying to implement.
1. Create a burning platform for the change, in terms that are locally relevant.
2. Create a coalition for change (typically executive council or steering team).
3. Establish the vision for what change will look like and what will likely happen if the change is not implemented.
4. Communicate that vision, again in terms that are locally relevant.
5. Empower individuals to act.
6. Attain early wins (pick easy wins first so as to build confidence and show results).
7. Build on momentum by increasing the pace of change (build on the flywheel effect).
8. Anchor the change in disciplined standard work.

Bob Dodge: I use a blending of LaMarsh, Prosci, and my own experience, which is to know no one mythology will be acceptable in such a diverse world. There is too much experience with so many approaches. I know I must work with what is in place that works and help my clients apply methods to improve those areas that need improvement. I am typically working with much smaller firms these days, and there is no capacity for the rigor of a brand name methodology. I must work with the business owners to help them: Define the change, determine who might resist and why, and mitigate that resistance.

David Ringel: Kotter, as a baseline method/guiding principles

Joe Beggiani: LaMarsh—it is what I was trained in. But from my experience they all are very close on what needs to be done to achieve a strong outcome.

Lynn Doupsas: Kotter and Procsi methodologies are most common. I use a combination or tools, techniques. Also, good standard project/program management techniques.

Gary Bradt: John Kotter's eight steps is my favorite methodology for making change happen. It's been around for a long time, and it's been tried and tested. It does lean more on the leadership side, but there are some change management parts of it, as well.

Roxanne Britland: I've only practiced in the LaMarsh model. That's why I use the phrase "managing change" a lot.

Darryl Bonadio: Yes, AIM.

Bob Von Der Linn: GE CAP: The GE approach has always been to put simple, effective tools in the hands of all managers, and teach them how to use the tools. This approach makes them better leaders, and it enables change agility. It works.

Marc Zaban Fogel: John Kotter's work. Switch, Heath Brothers.

Currin Cooper: I use a hybrid approach. I believe that change is a process, not an event.

I have used Conner Methodology, PROSCI, Accenture, Kotter, ACMP, and others. The basic steps are the same:

- Understand and articulate the intent of the project. There needs to be a crisp summary and a common vision. People need to be able to grasp what the end result will be, without having to read the project charter. The people impacted need to understand where they fit into the process.
- Identify and understand your stakeholders. What do they need? What kind of information, training, skills, or preparation?
- What are the gaps from the "as is" and the "to be"? What is the delta?
- What are the risks to the project that won't be managed by the project manager? Often in change management, the "out of scope" items are the ones that bite the hardest.
- Communicate, communicate, communicate. Use as many channels as possible. Use as many existing, or "trusted," sources as possible. Use them in person, when possible. Make sure everyone involved on the project team can verbalize the intent (see first

bullet, above) and knows where to direct the Chicken Littles ("The sky is falling!") and the "old schoolers," doggedly fixated on continuing to do their wonky, archaic process.
- Will it stick? Did the change do what it set out to do? What are your measures of success and can you measure them? What does success look like? Is work productive?

Mike Ensby: Not really. Interesting that a search of "X" Management models reveals over 200 CM methodologies. The best thing is to borrow liberally from several models and develop a robust heuristic that can be situationally applied. Kotter is good, but so are the many variations. Avoid becoming a "one-book wonder."

Patra Madden: ADKAR (Prosci model). The assessments and tools that they have developed are excellent for technical people to grasp. (I'm an engineer and work in a very technical company.)

Scott McAllister: I grew up on the Kotter approach, and since getting more exposure into the Prosci model, I really appreciate the Prosci approach, because it connects individual and organizational change. I learned from John Kotter that the first step is Sense of Urgency, and he's written in the Harvard Business Review that even if you don't have it, sometimes you need to fabricate it. Not all change will benefit from approach. If you're upgrading [the] Windows operating system, fabricating a sense of urgency isn't going to be the path to success. I really appreciate that the Prosci approach works on lots of different kinds of changes and is designed to be sized and scaled appropriately.

Rick Rothermel: I am compelled to recommend the LaMarsh Global Managed Change Methodology. Our clients tell us they like the approach because it is simple, scalable, easy to understand and apply, and results in data-driven decisions and outcomes.

Michele Quinn: ADKAR Model – Prosci 3 Phase Methodology

Jessica Bronzert: I'm trained and certified in the Conner methodology. That said, there are many methodologies on the market, but they are not substantially different from each other. The key factor is having a seasoned change practitioner doing the work. The best practitioner can leverage a basic methodology and have a positive impact; however, an inexperienced practitioner won't have the presence to influence sponsors and other stakeholders, no matter how good his/her tools are.

Siobhan Pandya: No favorite—fit for purpose is more important to me.

Q14. Key metrics? How do you measure the effectiveness of change management?

Tom Cluley: I use a proprietary assessment tool that measures the elements I defined in the critical success factors. It relates to an "S" curve that defines where an organization is on its transformation curve.

Bob Dodge: This has always been a difficult one, and in my current world, even more challenging to come up with quantifiable metrics. All I can go with are more behavioral:
- Do people understand the *why* and the *what* for their changes?
- Are we sensitive to their difficulties with the *how*?
- Are we identifying why they might resist?
- Are we successful in reducing that resistance?

David Ringel: Percentage of projects that result in an implementation of a change, percentage of improvement of project metrics, percentage of monitored risks realized (and percentage of those that were successfully overcome), trending rate of change (project frequency).

Joe Beggiani: There are all kinds of metrics. Below are just a few that are very good at measuring change.
- Employee satisfaction survey
- Adoption metrics
- Employee readiness assessment results
- Proficiency assessment results
- Usage reports

Lynn Doupsas:
- A: Rate of adoption (systems)
- B: Employee engagement scores
- C: Customer satisfaction scores

Scott Leek:
- Made the change (yes/no)
- Project goals met (e.g., reduced cycle time by 30%)
- Climate survey of population ("How did it *feel*… ?")
- Expert panel autopsy and lessons learned

Sarah Carleton: The ultimate (though lagging) metric is the success of the project. Leading metrics include survey scores of team members, stakeholders, process owners, and sponsors; time spent addressing change management aspects; risk management scores.

Roxanne Britland: By mitigating resistance—no stalls. No pauses. By actually getting to the results and making it all the way through implementation.

Darryl Bonadio: Observation of the necessary behaviors.

Bob Von Der Linn: The best measure of change effectiveness is ROI for the project. Some leading indicators would be percent and speed of adoption, participation in training, communications milestones on time. I have used employee surveys effectively to gauge effectiveness of change strategies. It is very hard to quantify.

Beth Cudney: Absenteeism. If the culture of the organization is strong, then employees will come to work, even if they are a bit under the weather.

Currin Cooper:

For data: You can do a before/after survey. You can measure support calls on "how to" questions. You can measure how long a process takes, how many errors, typical Six Sigma stuff. Training surveys can measure understanding of the new way of working or where folks are on the change curve. If training was inadequate, there will be errors. You can measure turnover, and you can measure your ability to fill vacant roles.

Bottom line: Who's hair is on fire on day 1/cycle 1? Whose phones are blowing up? If things are pretty quiet, it means you did a great job with change management. People may not like the new process or environment, but it means they're trying to work it out.

I have learned that the key success criteria take months or years to manifest. If you have to back out and go back to the old process, do a lot of rework, or overhaul workflow, it was not a success. But, this is slow, and usually the books are closed by that time.

Here is an example of where investing in change management could have saved a lot of time, money, and heartburn:

A very large public school system rolled out a Pay-for-Performance program. They spent millions collecting metrics and building databases and dashboards. They trained the Central Office. They promoted the new approach.

No one who was directly impacted wanted it. The teachers didn't want it, the principles sure didn't want it, the parents didn't want it, and the students just wound up spending more time taking tests to measure their teacher's performance, so they didn't want it.

    Guess what? The superintendent left the district before the directive was fully implemented, many teachers quit, and ultimately the pay-for-performance initiative was repealed.

Mike Ensby: Stakeholder expectations surveys, and future performance gains. Better gains usually mean better CM structures during the project.

Patra Madden: Very difficult. My biggest metric is talking to people. Unorthodox, I know, but it gives you a good idea.

Scott McAllister: Adoption and usage of the solution. The nice thing is that ADKAR provides a rich set of adoption metrics in and of itself.

Abdiel Salas: Results versus time.

Rick Rothermel: Did the organization achieve the intended results in the time frame, to the level of quality and the investment allocated?

Did the organization achieve an "acceptable level of acceptance"—have those impacted by the change accepted and adopted the behaviors required for the organization to continue to achieve the performance levels set by leadership?

Michele Quinn: Speed of adoption, ultimate utilization, proficiency.

Jessica Bronzert: There are three major areas of measurement for effective change management: (1) risks, (2) commitment, and (3) realization. There are known patterns of risk that contribute to the success or failure of transformational change. Monitoring and mitigating against these risks helps make the change more successful. Leader and employee commitment to change is critical to its success as well, and this can be measured. Lastly, realization is the delivery of the original business outcomes that got the change approved and initiated in the first place. Most change management efforts stop at installation—when the system is turned on or the people are trained—and doesn't continue until the business benefits of that system or training are realized.

Siobhan Pandya: Business and performance targets, employee engagement scores, feedback from internal communities, number of visual metrics, recognitions awarded, CI projects running and completed, number of engagement sessions, number of success stories.

Q15. Please describe the biggest mistake you've made (or seen being made) with regard to employing change management.

Tom Cluley: The biggest mistake I've made is not getting leadership engaged and driving the change. The second is not taking baseline data up front, so as to measure the impact of the change.

Bob Dodge: I was sent to a client to assist in implementing change management, my way. They did not know our Methodology, but insisted on implementing what they envisioned would help them. It became a point of contention; so much so that I was asked to let someone else help. That person could not implement the solution we had in mind any better than I did. The lesson for me was to be clear on what the desired solution is and go with what they want initially, and move them incrementally to what works, as we both discover the flaws in their vision of success.

David Ringel: Doing it in an overt, obvious, in-your-face manner. Better to be subtle and organic/vested.

Joe Beggiani:
- Cutting corners on the process
- Running change management separate from the project/program
- Change effort not properly funded
- Sponsor delegating responsibility
- No reward system

Lynn Doupsas: Usually it is around not garnering the right executive level support.

Scott Leek:

Have seen versions of the following on multiple occasions:

Patronizing and insincere attempts to persuade target population that the answer being pushed was not only right, but the best; not listening to legitimate concerns raised as part of the change management process, ignored or dismissed; and, generally coming off as "… we know best, we just have to communicate better so those of you who don't 'get it' will finally understand…"

I call this one-way change management… because it is only going to end one way… bad.

Gary Bradt: Leadership not understanding the needs of the people impacted by the change and just telling them they need to go change without explaining why. Along with that, if there's going to be some pain involved in that change, why do we need to go through this pain? And the biggest mistake I'll see in the same vein is, "well here's how the company is going to benefit, here's how our stakeholders are going to benefit, here's how our customers are going to benefit," but if you can't explain to people how *they're* going to benefit somehow, someway, you'll lose them.

Sarah Carleton: The biggest mistake I've made is neglecting significant stakeholders who came back to haunt one of my LSS projects.

Roxanne Britland: Resistance is such an important element of change management. Leaders are entitled to ask whatever questions they want, and just because they're asking a question doesn't mean there is resistance. It may mean they simply don't know. Sometimes we're focused on the idea of resistance a little too much. It makes you feel negative.

Darryl Bonadio: The biggest mistake I have ever seen is acting like the change was good for everyone. Sometimes changes are not good for people, but are still necessary. People are not stupid and will see through this quickly, and resistance will be extreme.

Bob Von Der Linn: Aside from the obvious big ones (e.g., not using it at all or engaging CM resources too late), the two biggest problems I see are (1) Not having an organizational structure that can integrate and leverage multiple change initiatives, and (2) Not having an effective means of determining the level of CM support an initiative requires. Not all changes are equal. Budget and size of impacted population are unreliable predictors of need for CM.

Marc Zaban Fogel: Not being direct with a leader who wasn't committed the full scope of the change he/she was asking people to make.

Beth Cudney: The biggest mistake that I have seen made is training individuals on Lean Six Sigma without change management.

Mike Ensby: Sticking to the "model" when the mood says otherwise. Better to have an adaptable heuristic. Having a technically competent, but organizationally clueless project lead who is bound and determined to drive the clown car over the cliff.

Patra Madden: Stopping after the change is implemented. Implementation is not the end of CM focus, but the beginning of the more difficult CM focus. Not the time to stop!

Scott McAllister: Starting with a bunch of training and thinking that's going to solve all of the problems.

Abdiel Salas: Assume everybody sees or understands everything at the same level as you do.

Rick Rothermel: Many practitioners interpret change management as the application of tools and methodology, failing to recognize that the purpose of tools and methodology is to provide a framework for analysis, decision making, and resistance mitigation activities. I have also observed emerging practitioners over-complicating the application of the methodology. I have operated throughout my consulting career from the philosophy that the amount of change management methodology and discipline required is dictated by the specifics of the change and the level of resistance that requires mitigation.

Michele Quinn: Leadership believing that the "Right Answer is enough." You can have the best solution, but without adoption and usage by the people impacted by the change, the return on the investment will be limited.

Jessica Bronzert: When the change management function itself is not sponsored well enough to partner with the business making changes, it will be difficult, if not impossible, for change practitioners to be successful.

Siobhan Pandya: Lack of communication, including lack of clear communication and message and also lack of follow-up after communications—false sense of confidence that message is understood and all are in support.

Q16. Do all Lean Six Sigma projects need change management? If not, how do you know when it's needed?

Tom Cluley: They all need some level of change management. Even a small project on a production line requires the workers to accept the change.

Bob Dodge: Absolutely. It is needed in every case, now. If we do what I described above [in Question 15], and the answers are yes, there is not much more to do.
- Do people understand the **why** and the **what** for their changes?
- Are we sensitive to their difficulties with the **how**?
- Are we identifying why they might resist?
- Are we successful in reducing that resistance?

If people are genuinely willing to leave the current state, have bought in to the desired state, can and will go through the delta, with no issues over sponsorship, history, or their own wiring, we are good to go. If we are going to invest in the resources to "do" Six Sigma or Lean" why not make sure people will make the change?

David Ringel: Yes, all.

Joe Beggiani: Tricky questions… my answer is yes, whenever a process, design, build, or whatever the change might be, the communication to the effected group of employees still needs to be done. The change could mean new training do to the change of skills, or the understanding of role changes, and even the reduction of needed personnel. Change management can elevate a number of risks that any size organization needs to build adoption to stay on top of the game and remain productive.

Scott Leek: Yes. When something has to change, and that change will impact multiple people, change management should be used. The question is really one of degree of formality and detail.

Sarah Carleton: Change management is an integral aspect of all projects.

Roxanne Britland: As long as the solution is introducing a change in the business, in terms of somebody's role, a process step change, then change management is needed at that level. As the solution becomes more complex, and as it affects more people, change management is more likely to be needed.

Darryl Bonadio: Do they all need the "same level"? No. But, all Six Sigma projects are about change, so using change management techniques and methods make sense.

Bob Von Der Linn: Ask yourself, "Will this project change how people do their jobs, and if so, how much will it change?"

If it is a developmental change, i.e., moving to a clearly defined future state (e.g., transitioning from an existing software application to a different, new, cloud-based application), by definition, you will have 100% adoption, but this population will still need communications, training, manager preparation, support, etc.

If it is a transformational change, i.e., moving to an undefined future state (e.g., reorganizing around a new operating model, off-shoring some work, and introducing new tools and processes, all simultaneously), now in addition to the typical support, you need a shared vision, transparency, employee input in process design, updated measurements, rewards, training, etc.—all change management practices.

Marc Zaban Fogel: Any project that changes the way other people work in a way that is different and/or requires more effort or decision making, should include some thinking on change management.

Interviews with Experts • 139

Beth Cudney: Yes, all projects need some level of change management. BBs and the Lean Six Sigma team will be working with individuals on making changes. No matter how small the change, if it is not handled appropriately, it can negatively impact the success of the project or the culture in the organization. Change management can be as little as how to inform people about the project. An authoritarian approach to certain individuals can immediately make them resist changes associate with the project. Therefore, it must be addressed in every project.

Currin Cooper: Sometimes training and additional information is all that is needed. Good project managers intuitively keep an eye out for change management pitfalls.

Mike Ensby: All do. It's a matter of scope and scale. It's not a true LSS project if it doesn't result in change.

Patra Madden: YES!!!

Scott McAllister: I'd say the vast majority need change management. And, how do you know when change management is needed? You have people in the organization that need to change how they do their work. If yes, you need change management.

Rick Rothermel: Every project requires a level of change management, because no matter how large or small the scale of the change, if people are impacted, their resistance needs to be acknowledged and mitigated.

Michele Quinn: Yes, but the level of time and resources applied needs to be evaluated and determined, using proven risk assessment tools and organizational readiness tools.

Jessica Bronzert: Engage change management practitioners at the beginning of any ideation around projects. Seasoned practitioners have approaches to determine if the disruption from an LSS project requires change management, and if so, what level of support.

Siobhan Pandya: Yes.

Q17. How does a BB (the project leader) know when it's time to ask for help with a change management problem?

Tom Cluley: When they can't overcome resistance to change within the workgroup or leadership members of the client.

Bob Dodge: If the BB cannot measure the risk, or the Risk Management Assessment indicates the risk is high, help is needed to make that assessment or to mitigate the risk. If it is important enough to determine what to change, it is important enough to understand why the change may fail.

David Ringel: When they sense there is no progress or momentum has stalled.

If a stakeholder is beginning to lobby others against the change.

If you start to hear "noise" from people who have not been directly involved.

Stakeholders sustain or increase levels of argument, quietness or challenge.

Joe Beggiani: If they bring in the change manager in the planning stage, there should not be a problem. But, all too often, the project leader does not pull in a change manager until communication leaks, employee resistance, and bad water-cooler talk has already begun. This should not stop the project lead from engaging the change team to get involved.

Scott Leek: Change management is a process of asking for help. If they are doing change management, they are asking for help by definition.

Gary Bradt: When you can't sleep at night.

Sarah Carleton: BBs should ask for help from their team at the very start of their projects. If there are any significant roadblocks, they should ask for help from their mentor.

Roxanne Britland: When there is resistance to the solution.

Darryl Bonadio: When they recognize that there are issues with either sponsorship or resistance from the targets of the change. But, the best answer is that unless they are formally trained in change management, they should always seek advice to make sure they are setting themselves up for success.

Bob Von Der Linn: When a project charter is being drafted. ☺

Marc Zaban Fogel: When they know a key person or group is negative or neutral on the change.

Beth Cudney: A BB should work with the project sponsor and champion before the project starts to understand potential resistance and should keep them informed throughout the project, so that potential issues can be identified before they happen.

Currin Cooper: A key stakeholder group is missed. Stakeholders are confused and frustrated. Damage control is taking up a lot of time. Rumors are out of control.

Mike Ensby: When the level of commitment required doesn't materialize, and the team members start leveraging competing priorities to avoid the tasks at hand.

Patra Madden: I insist that all my BB become CM practitioners (Prosci), so we never need to ask for help. And, we know when to recommend to a sponsor to have a practitioner assigned to the follow-on with the team.

Scott McAllister: The CAP program talks about quality × acceptance = Effectiveness, and when you are only focusing on the quality and not the acceptance, you're never going to be as effective as you want to be. So, I'd say a BB or a project leader should ask for help when they start seeing resistance from employees ... when the people side of the equation starts to be more challenging than the technical side of the equation.

Abdiel Salas: When projects become longer than expected, i.e., 6 or 12 months. A BB should be an expert in change, and as experts, we need to understand the company's background on change. There is a certain amount of change a person or department can manage. As a BB, you should find out.

Rick Rothermel: They shouldn't wait to ask for help. They should incorporate change management methodology and discipline into the project plan from the beginning. They should acknowledge that the people impacted by the change will have issues, challenges, as well as opportunities, and that a change management lens will improve the probability of success.

Michele Quinn: Start with a Change Management strategy as an integrated component of your project plan; be flexible, adjust as necessary.

Jessica Bronzert: When they first get wind of the project! Change management is most effective when considered up front—even before a project is approved. Indeed, change management as a discipline has a lot to offer in establishing the intent of a change and helping leaders determine if they are truly committed to seeing it through.

Siobhan Pandya: When you are not progressing with the stakeholders and the project—you may not have everyone where you want them to be or the project complete, but they are moving in the right direction. If you are not moving at all, or if blockers are being intentionally created, then it is a good idea to request help.

Q18. How can you measure the level of resistance to a project?

Tom Cluley: Through team participation and seeing and listening.

Bob Dodge: Aside from asking questions, surveys, and observations, it boils down to determining:

- Do people understand the **why** and the **what** of their changes?
- Are we sensitive to their difficulties with the **how**?
- Are we identifying why they might resist?
- Are we successful in reducing that resistance?

David Ringel: Progress

Joe Beggiani: There are tools that you can use, such as a history assessment of employees impacted by a previous change, culture assessment of the impacted employees, identifying the root causes of resistance, engaging the resistance to gain feedback and acceptability.

Lynn Doupsas: I would power map all the major players and rank them green, yellow, or red in terms of their level of support.

Scott Leek:
- Assume it is not zero, and tap into the informal network of chatter.
- Target thought leaders for direct or indirect questioning.
- Talk with the people who talk with the people.
- Ask.

Gary Bradt: I don't know if you measure that as much as you feel it. You need to be out walking around. You need to be walking the halls and talking to people. You need to invite dissent. You need to say, "Hey, how's it going?" and "How do you feel about this change?" and "What worries you the most about it?" It needs to be more of a feel process. You can measure things like how many people are showing up at meetings and how many are meeting their deliverables and things like that, but I think resistance is more of a heart thing—not a head thing. You've got to touch it and feel it. You've got to talk to people. And, if there is resistance, rather than telling them that they're wrong, try to understand why they feel the way they do and see if you can address what's behind the resistance.

Sara Carleton: You can measure resistance with survey scores and with delays in accomplishing project deliverables.

Roxanne Britland: When you scope a project, you identify all of the stakeholders who can influence that solution or are impacted by the solution. You then have to build a relationship with all of those people. Resistance comes from people, not from anywhere else.

Darryl Bonadio: There are several factors to consider. First is the culture of the organization and its acceptance of change. Next is the support from impacted parts of the organization. This is often done through interviews or tribal knowledge. We highly recommend that all improvement projects do a stakeholder analysis that sizes up key individuals or groups and their current level of support for the change. Then, a communications plan can be developed to address or mitigate resistance.

Bob Von Der Linn: I gave up trying to quantify resistance a long time ago. I assume that there will always be some level of resistance and that we should plan for it. Having said that, if employee engagement scores are low—indicating a poor relationship between leaders and the employees—there will be resistance. In my experience, the most significant predictor of resistance is the "capacity" of the affected organization. If they are already overworked/understaffed and/or change fatigued, because they've been in turmoil for a while, you can count on significant resistance. If there has been a change of leadership and/or reorganization, things tend to move slower due to lack of trust and connection—that can appear as resistance. If you have metrics and incentive systems in conflict with a change, you will definitely have resistance.

Marc Zaban Fogel: The speed of getting basic information and feedback from stakeholders.

Beth Cudney: It is difficult to measure. There are signs, such as team members coming late to meetings, difficulty in getting people involved, and lack of ideas being shared during brainstorming.

Currin Cooper: Ask. Do an assessment. Interview people. Conduct focus groups. Have business representatives sit on the project team. Go out into the field and talk to people. Try to understand what they have today and what they like about it. Try not to break what's good about the current state.

There are plenty of assessments. I like the 4-square or the 9-square ones the best. They are accurate enough to tell you what you need to know, and take a lot less time and seem less bureaucratic.

Mike Ensby: The typical conflict warning signs: avoidance, arguments, increased level of CYA.

Patra Madden: Speed of adoption and ultimate utilization. I had a project that a lot of the stakeholders/users worked to develop. Everything they wanted was implemented, but when the tool was rolled out, the utilization was abysmal. And no one would explain why they weren't using it when asked. I decided to pull all 65 of the users together in a room and ran a "resistance workshop." (They didn't know I called it that.) I gave all of them a pile of Post-its and asked four simple questions: What do you like about the process? What questions do you have for the developers? What suggestions do you have for the developers? What don't you like about the process? After 15 minutes, the corners were full of Post-its. Lots of good suggestions, good questions, and about 60 Post-its all saying the same thing under the "What I don't like" page. They didn't want to be the one person to say it, but were able to get their voice heard loud and clear when they were speaking anonymously. The sponsor never would have guessed where the resistance would surface and didn't understand where the people were coming from. But he didn't need to understand it. He just needed to fix it. Which he did.

Scott McAllister: You can use frameworks like ADKAR to help identify and mitigate resistance… and to measure it to some extent. You can measure it by using stakeholder management tools, stakeholder analysis where you plot them on a spectrum on level of support.

Abdiel Salas: Time versus difficulty.

Rick Rothermel: Data! Talk with those impacted. Observe their behaviors and body language. Surveys, focus groups, interviews, and observations are all typical strategies for identifying resistance that can be translated into measurable data. If you have done everything possible to successfully install/implement the solution, it is likely that the real issue is the resistance factor of those who are impacted by the change, but still are resisting.

Michele Quinn: Observe behaviors. Listen to the questions being asked. Resistance is an indication that individuals or groups have an unmet need. I don't know that I would spend my time measuring resistance. Identify it, work to understand it, and then take action. I'd suggest investing time and effort in a solid, proven change management process, since research shows that 50% of resistance by employees and middle management can be avoided. First, resistance prevention by implementing effective change management, then proactive resistance management—anticipate it, put special tactics in place to break it down, and finally, reactivate resistance management as necessary, focusing in on observed behaviors, potential reasons for the resistance, and selecting and sequencing specific tactics to overcome.

Siobhan Pandya: Everything, from attendance at meetings, responses to e-mails, and provision of information, to actual project deliverables not being delivered.

Q19. What is the project champion's role (some call it the sponsor) with regard to change management?

Tom Cluley: To go beyond support to engagement. Actively participate in report outs, asking questions that challenge the team. The Change Manager should participate in what I refer to as GBUs (good, bad, ugly meetings) with the facilitator, to discuss what is going well, areas of concern that are being worked, and showstoppers that require management intervention.

Bob Dodge: The sponsor must be consistent in displaying his or her commitment and passion for the change, program, project, or change, while being just as diligent in observing why people might resist and helping them reduce the reasons for resistance.

David Ringel: Sponsors call for change, charter the project, and communicate with stakeholders that are considered key to the success of an event. They ensure resources are made available, roadblocks to progress are removed, and decisions to support the recommendations of the team are made.

Champions act as agents of the sponsor and key stakeholders with event team members and other stakeholders across the enterprise. They ensure communication and understanding are maintained throughout the life cycle of the event and interact with the team on a regular basis, advocating as required with the sponsor and key stakeholders on the team's behalf. This dual advocacy role of the champion can be summarized as the change agent for the project.

Joe Beggiani: *Sponsors must understand the change, meaning they need to define the desired state both numerically and in behavioral terms. They need to identify and empower change agents. They will also need to identify the amount of and type of loss tolerable in the delta phase. They need to engage senior management to gain adoption.

The role of the project champion is to identify a project's strategic objectives, ensure the vision for the project is successfully

---

* Taken from LaMarsh Change Master text.

communicated into requirements and design, remove roadblocks for the project team, work with the project manager to prioritize the allocating and organizing of internal resources.

Lynn Doupsas: Very important to help align and set the tone. Champion needs to be visible and active as well.

Scott Leek:
1. Help the BB understand the concerns and interests of the various stakeholders and develop strategies and tactics to obtain the necessary buy-in.
2. Assist the project team [in] breaking down barriers and resistance.
3. Help overcome cross-functional turf battles.
4. Lend political capital to gain support.

Gary Bradt: Tear down barriers and obstacles that the team needs removed, so as to be successful. The senior project sponsor's job is to be a servant to the people on the ground, making the change happen and saying, "What resources do you need?", "What obstacles can I help you overcome?", "What doors do you need me to help open for you?" So, in my mind, they're in very much a servant-leadership role. Obviously, somewhat of a cheerleading role as well, but making sure the team is executing the project and has the resources that they need to be successful.

Sarah Carleton: The sponsor should insist on using change management as an integral aspect of the project, help identify key stakeholders, and ask questions about change management during project reviews.

Roxanne Britland: Awareness of the project, endorsement and approval of the scope, engagement, commitment, and inspiration to the team members, then they need to walk the walk. It can't be lip synch. They really have to commit to team members first and foremost—that the right people are on the team who can contribute to building the solution.

Darryl Bonadio: They both need to communicate and demonstrate support. Their roles are to address barriers to change that arise during the project. The sponsor (in our world) owns the project and the results. The champion is higher up and can help by cross communicating to other functions. (We call everyone a sponsor, but differentiate the "project sponsor" from general sponsorship.)

Bob Von Der Linn: A great question—lots of debate here! A business leader is accountable for the success of their business unit. A project team is the leader's tool to implement a change. Ultimately, they are the leader of any change initiative. Change management is a core leadership competency; it cannot be outsourced to someone else. In 2016, if you cannot effectively lead your organization through change, you are not a competent leader and should not be rated or compensated as one. It is perfectly legitimate for a leader to enlist the support of a change consultant to coach them through a change, but to delegate change management to a hired gun in the PMO is abdication, pure and simple. When I learned CAP, we were taught to address this issue with sponsors at the project outset—If you cannot commit to visible, engaged leadership of this project (to the project team, and all the stakeholders), that is a red flag and the project is at high risk of failure.

Marc Zaban Fogel: Champions should be aware of all critical stakeholders and their issues. They should be ready when resistance is met at their level.

Beth Cudney: The project sponsor and project champion should be able to anticipate resistance, since they are closely working with the individuals who will be affected by the change. They should help the BB develop a communication plan that adequately addresses these fears and eliminates roadblocks before they occur.

Currin Cooper: This is well documented in a lot of places. Someone has to want the project. Someone has to go to bat when there's resistance. Someone in authority has to put their foot down and say, "This is going to happen. These are my expectations."

Mike Ensby: They are accountable and need open and collaborative communications. Trust and respect for the process.

Patra Madden: The sponsor and champion are required to communicate the benefits of the change and ensure the management team is aligned to the change and fully supporting.

Scott McAllister: The project sponsor needs to be active and visible throughout the life of the change, they need to build a coalition of support amongst leaders in the business, and they need to communicate effectively.

Abdiel Salas: There are problems that should be resolved at certain levels of the organization. The BB should be aware, should recognize, and should ask for help in a timely manner.

Rick Rothermel: The issue here is having a common definition of project sponsor and project champion. There is one person in the organization to whom all of the people impacted by the change ultimately reports. This person is the only leader who can hold the entire organization accountable for making change happen. His/her role is to hold everyone in the leadership cascade accountable for leading the change through his/her organization. Sponsors at every level need to understand their roles and responsibilities: Be actively engaged in and support the change, encourage those impacted to voice their issues and concerns, look for and leverage opportunities to reinforce and recognize individuals who are attempting the live in the "future state," discipline/inform those who choose to change the consequences and then enforce them.

Michele Quinn: Provide the business *why* behind the change. Build a coalition of support of other leaders. Communicate, communicate, communicate (talk and listen).

Jessica Bronzert: One of the misconceptions about change management is that it can be "farmed out" to project teams or change management practitioners, letting leaders "off the hook." Nothing could be further from the truth. While change practitioners can educate and support, only those who control communications and consequences (positive and negative) for those who must change can truly make a change successful. Indeed, the lack of effective sponsorship is the number one killer of transformational change efforts.

Siobhan Pandya: Provide a fresh perspective. Provide challenge if necessary. Assist in development of way forward. More coaching and teaching than just doing. Help to overcome blockers. Ensure that progress is on track. Update on any strategic changes that may impact the goal.

Q20. Are gaining adoption and eliminating resistance the same thing? If not, please explain the difference?

Tom Cluley: Not really. Adoption can be gained through authoritative oversight. If resistance is eliminated, [people] will accept the results without oversight.

Bob Dodge: The difference is in "forcing adoption" in spite of resistance. We might achieve adoption, but leave a lot of bodies in the wake, create far too much stress, add to a history of forced changes, and risk reversion to the old ways if we don't address the root causes of resistance. "It's my way or the high way" will get adoption, but at what cost?

Interviews with Experts • 149

David Ringel: Gaining adoption can be accomplished on any project through consensus building (effectively mitigating the effect of the resistance). Eliminating resistance would require the source(s)/cause(s) of resistance to be eliminated. If the cause(s) remain, the resistance will surface again and require gaining adoption.

Joe Beggiani: They are not [the same]. Ggaining adoption can have a number of reasons: high tolerance to change, open-minded employees who thrive on challenges, employees who have access to the right information due to good communications from management, as well as employees who were involved with user testing or those employees that pushed for change to better the organization.

Eliminating resistance is done by doing the homework on the culture, history, and resistance of the targeted group, as well as laying out a proper communication plan, and executing that plan. Acquiring senior leadership buy-in early and keeping them in the loop on the change throughout the change life cycle is key.

Scott Leek: They are not the same. Eliminating resistance is eliminating a negative. Gaining adoption is affirming a positive. Eliminating resistance makes you no better off; it makes you no worse off. Gaining adoption makes you better off, because that means the object of the exercise is occurring, namely, change.

Gary Bradt: Gaining adoption is on a continuum. You'll get people who will say, "OK, I guess I'll do this, I guess I'll go along, we'll see what happens." At the other end of the continuum, you have people who'll say, "This is the greatest thing ever. What can I do to make this work? I can't wait! I'll work morning, noon, and night, because this is so exciting." So, gaining adoption has a continuum.

Overcoming resistance kind of gets you to the point where people say, "OK, I'll stop resisting you here," but where they are on the continuum is really what matters. They're two different things. You've got to overcome resistance before you can get them to adopt. And then, when you get them to adopt, it's about where they will be on that continuum.

Sarah Carleton: Gaining adoption is more inclusive than just eliminating resistance. Gaining adoption may include eliminating resistance, overcoming inertia, increasing motivation, providing tools and resources, highlighting good examples, rewarding appropriate behaviors, and recognizing improvement.

Darryl Bonadio: I think you are asking is there a difference between installing the change and truly implementing the change. Yes, the latter is changing behaviors. If you do not see the new behaviors, you probably just installed the change, and it will not hold.

Bob Von Der Linn: They are not the same thing. Adoption is the desired outcome. Eliminating resistance is a part of your change strategy to achieve adoption.

Marc Zaban Fogel: They are two sides of the same coin.

Beth Cudney: Eliminating resistance is proactive. It is predicting potential resistance and eliminating it before it can occur through strategies such as communication. Gaining adoption is reactive. It is after the change has occurred and getting individuals to accept the change.

Currin Cooper: This is a good question. They are related. However, not adopting the change is not the same thing as resisting the change. Resisters want the change to fail. They want the change to go away so they can keep doing what they are familiar [with] doing.

Both require enforcement and consequences to doing work the new way. If the consequences aren't important to them, if they are too far into the future, or if they are uncertain what's expected, you probably won't get adoption from the folks on the fence or the resisters.

If there is no governance, or the governance isn't enforced fairly, resistance will grow.

Mike Ensby: No. People will adopt but not adapt to the changed state (participation over commitment), if they are resistant to change.

Patra Madden: My story about the resistance came about, because I noticed the lack of adoption. The reason for the lack of adoption surfaced by delving into the resistance. They are linked, and one leads to the other.

Scott McAllister: I wouldn't say they're the same thing. They are both realities, so in some situations, you've got to drive adoption, while in others you might have a change that is experiencing lots of resistance, and so you take different steps to address both desired outcomes. They're definitely connected. Often-times people talk about change management as managing resistance. That's a part of it, but not all of it.

Abdiel Salas: They are different. People first accept the leader and then the idea. Resistance has to do with knowledge, while adoption has to do with process change.

Rick Rothermel: Resistance represents the issues and concerns from those impacted by the change that get in the way of the individual making an informed decision to adopt or not [to] adopt the change. All resistance could be eliminated, and yet an individual may still choose not to adopt. Adoption happens when the individual accepts the new behaviors required and demonstrates them in his/her actions, regardless of whether resistance is fully mitigated.

Michele Quinn: I think of it in terms of compliance versus commitment. Mandated change can result in compliance, but for real, long-term, sustainable change, commitment is required.

Jessica Bronzert: Yes and no. It's tempting to simplify commitment and resistance as two sides of the same coin, but there are different models and ways of thinking about both concepts. Resistance and commitment are complex ideas that have many layers and nuances underneath. Different models will speak to the emotional experience of change, the behavioral or observable experience, and the mental experience.

Q21. With regard to reducing resistance, how would you address a group of "thinkers," i.e., highly analytical, data-oriented people? How would you address a group of "feelers," i.e., people who are not interested in facts and data?

Tom Cluley: With thinkers, stick to the facts regarding the need to change, the positive impact of that change, and the detrimental results that will occur if the change is not implemented.

With feelers, discuss the positive social impact of the change and how it will affect people in a positive manner, as opposed to how not changing will negatively affect people.

I like to use Clare Grave's values systems, where you address differently the values systems of individuals that are Tribal, Egocentric, Conformist, Competitive, Sociocentric, or Existential. When talking to a group, I try to address all of the values systems. When talking to an individual, I try to address what I perceive their value system to be.

Bob Dodge: The analysts need to understand the significance of the change—the *why* and the urgency of the change. They need to be brought up a level to understand the need to move quickly. The feelers need to understand the need for structure and the lack of certainty without the detail work. Gut feel and feeling good alone do not assure success.

David Ringel:
>Thinkers: Talk about the current state, gap, and potential results in tangible measurable terms, as well as the association between those measurable terms and the performance of the organization (or system) under review.
>
>Feelers: Talk about the current state, gap, and potential benefits in terms that align with the people factors (quality of life [QOL], quality of work life [QOWL]) and the association of the people factors and the performance of the organization (or system) under review.

Joe Beggiani: For analytical thinkers, I would use data to help drive key points and help produce conversation. For those who are feelers, I would still show some data, but have open dialogue and gain discussion through Q&A and feedback sessions.

Lynn Doupsas: This is why I became MBTI certified: to help teams understand differing approaches to dealing with change and communication. This is why I would prefer, for instance, to talk about something over the phone (as an extraverted feeler) than fill out a questionnaire. Additionally, I would say I practice the *Art* of change management not the *Science* (which is more of a BB/PMI trait) because I have high gut intuition.

>Change, adoption, resistance are human factors. In many ways it is the highly attuned person (in this case, [the] BB) who can accept what he/she may have a blind spot on and compensate by engaging a strong change management professional to assist. The most successful programs recognize and celebrate the need for a variety of expertise and communication styles and approaches.

Scott Leek: Seek first to understand, and go to where people are. Don't try to make them come to you.

Gary Bradt: You need to understand that you have to talk to them [the feelers] with *their* language, versus the people who are really into the details and the facts.

>When you're talking to the data people, give them the data, give them the information. When you're talking to the feelers, don't give them the information that would convince you. Give them the information that will convince them.

Sarah Carleton: Using the preferred mode of communication of the audience helps the message to be more impactful. Thinkers may like to see the data, while feelers may like to hear about how relationships are affected.

Roxanne Britland: With data-driven decisions being a theme, regardless of whether you're a data person or not, a compelling story is driven by good data. So, if you are in front of non-data people, you have to pull that compelling story out of the data and give it to them in non-data ways. I use data appropriately analyzed to build information that allows you to develop a compelling story.

Darryl Bonadio: The best way is to include people in the process. Mindsets and behaviors address both by educating people about the change and understanding the impact. No one likes to change, so understanding what is considered important to people will help identify communication methods that will be more successful. Will it be education, addressing fear or uncertainty, or gaining ownership of the change [that is] the best approach? Or all three?

Bob Von Der Linn: Every change strategy must address both the heart and the head to ensure it can reach all styles. An old adage from the art of persuasion is, "Never a statistic without a story, never a story without a statistic." I have used tools such as TRACOM® Social Styles assessment, or DiSC®, to help leaders and teams understand why this is important and to help develop robust change strategies.

Marc Zaban Fogel: Provide the case and ask for analysis and input from the thinkers. For the feelers, tell them the benefits and ask for their reactions or concerns and ways to address those concerns.

Beth Cudney: With a group of thinkers, I would show them the data to justify why the project or change is needed. For a group of feelers, I would make the business case based on what is happening within the organization and how it will benefit the employees. For an effective approach, both should be included.

Currin Cooper: They are not as different as you would think. If they communicate with you using PowerPoint, use PowerPoint. Give the analytical people more data, more spreadsheets, more charts, and more detail. Give the feelers the summary information, the big picture, bullets, analogies, and stay away from the data tables!

BEST: USE BOTH! Appeal to folks on multiple levels. Invite the loudest complainers to a session, and ascertain what the real problems are.

I did this once with a critical group of reporting power users (analytical, data-oriented), and I swear, the answers were all things like, "It won't work." When asked why, they couldn't articulate the issues, only that, "It's not what we're used to," or "It's different." (Wow.)

So, they vented (these were SQL developers). I documented their issues and created a table with a proposed solution/mitigation. I came up with the list of things to facilitate the transition, offered up tools, data mapping documents, training, workshops, and other resources. They felt heard, and I was able to run things by them, get them to pilot some solutions, and get their input. Input = buy-in.

The Go Live went great, by the way.

Mike Ensby:

Thinkers: logic is not common sense; while the root cause analysis may lead to a "natural" solution that may be unnatural to the folks who will bear the brunt of the change. End users are not always sufficiently engaged early enough.

Feelers: this is not about making people happy, it's about satisfying organizational needs. Status quo is more disruptive than structured disruption. Data is a good starting point to understanding "pain."

Patra Madden: I find Post-its work well for both types of people. The thinkers usually spend a lot of time sitting before they put down any words, and the feelers start writing the minute I say "go."

Scott McAllister: You've got to know your audience. With thinkers, you can use more of the analytics and data-driven decision making, and in the change management world, I think this is where you can get into the tracking metrics from an ADKAR perspective, where we often describe things with adjectives instead of numbers.

For the feelers, we may talk about what we are hearing, seeing, and feeling. These are sort of feedback elements, and we can then take an appropriate course of action. Instead of presenting graphs and numbers, it's often statements or verbatim from participants in the change journey that help with the feeler side of the equation.

Abdiel Salas: Balance is the word. All kinds of people are needed. Have a level of enough data, and use appropriate words to feed them. Become one of them, but challenge them to their level of commitment. Even thinkers are emotionals, and emotionals are thinkers at some degree.

Rick Rothermel: I would not make any distinction. The question asked is, what factors, issues, concerns, and challenges are getting in the way of your making the decision to embrace the change and adopt the required behaviors?

Michele Quinn: I use the DiSC model to understand my stakeholders, their motivators, fears, likes, dislikes, preferred communication mode, conflict mode, etc. There are three ways to engage people—logos, ethos, and pathos. The key is to understand your stakeholders' style and utilize the appropriate mix of each as you communicate the vision and address the questions, concerns, fears, etc.

For people not interested in facts, I'd focus on using examples, stories, testimonials; engaging their emotions. For the thinkers, I'd use logic, data, facts, research, etc. But a quick caution: no one is just one style. All of us are a mix of styles, so a balance of logos, ethos, and pathos is critical.

Jessica Bronzert: Resistance fundamentally comes down to questions around ability and willingness. Ability is about not only the skills or capabilities to be successful doing the new change, but also about my power, status, or position as well. Willingness is about my desire (or not) to expend the energy necessary to be successful with the change. Thinkers and feelers may articulate their ability or willingness concerns in different ways, which would guide a change practitioner or leader as to how to best address those concerns.

Siobhan Pandya: It is about balance. For the thinkers, data should be used to "get a seat at the table," to show that the data demonstrates there is need for this project, but also requesting more data from them. This is where they will thrive and be able to see from their own generated data that they need to move with others and reduce resistance. At the same time, you also need to introduce the feeling aspect to them, as it is not just about data and not just about feelings—balance is key. They will be more receptive due to the initial conversation and data support. Vice versa for the feelers.

Q22. Do organizations with excellent employee engagement still need change management?

Tom Cluley: Yes, because resistance to change doesn't just occur in the workforce, it also occurs in leadership. As noted in Kotter's 8-Step Process for *Leading Change*, communication and engagement are just a part of the equation to creating change acceptance.

Bob Dodge: Sure. It may be easier since they might be more willing to share their resistance than an organization of people afraid to engage, to "show up," or to collaborate to increase acceptance.

David Ringel: Yes.

Joe Beggiani: Yes, they are separate pieces in the company. Change management is a methodology to transition employee(s) and organizations, using a process to redirect modes of operation that significantly restructure a company or organization. Employee engagement is used to strengthen the culture and values of the corporation.

Lynn Doupsas: Yes, there is always room for improvement. It really helps if there is high employee engagement, and the key agents are helping to craft and disseminate the right messages.

Scott Leek: Yes, but not the same as organizations without it. For example, in organizations with excellent employee engagement, change management might focus more on change adoption versus eliminating resistance.

Gary Bradt: Of course! If an organization has excellent employee engagement, that's wonderful. Hopefully, that means they will be more open to whatever changes you need to introduce. The external world is changing so rapidly, if you're not constantly changing, you're falling behind, and you're going to die. So, whether you've got engaged employees or whether you've got walking zombies, you need change management.

Sarah Carleton: Yes, they probably are using change management to sustain employee engagement, because the world is changing at a rapid pace.

Roxanne Britland: I really believe that change management needs to be embedded innately into leadership and managers, and if it is, it's just the way you do business. You're always able to explain why we're doing something. You're always able to reward.

Darryl Bonadio: Yes, in fact sometimes more. Engaged employees take more ownership of the current way, so you need to make sure you address their needs and understanding for the change.

Bob Von Der Linn: Of course. Employee engagement is similar to trust in that it is fragile and can be broken. Fortunately, organizations with high engagement typically have an easier time managing change.

Beth Cudney: Yes, because you always need to have a communication plan and strategy to address change issues. By addressing them before they occur, you can mitigate risks.

Currin Cooper: I was on an employee engagement project. I still am not certain what that means. ☺

If they are engaged, it must be a great company to work for. They like things the way they are and may not want things to be disrupted. Change is disruption.

So, yes. They can be engaged, but do they know what's happening? Do they know how they'll be impacted? Will they lose their jobs? Will they be prepared? Will the change happen during their busy season? Will the training tell them how to do their jobs or just how to click through the system? What processes and controls will be in place to sustain the change?

Mike Ensby: Yes, even more so, because the expectations are increased for these types of organizations. The advantage is that the heuristic can be more adaptable, and you can look for opportunities to develop leaders at all levels of the organization.

Patra Madden: Absolutely.

Scott McAllister: Absolutely, I think they're closely connected. Organizations that have success with employee engagement tend to do a better job with the people side of change. Organizations that have poor employee engagement oftentimes under-serve the people side. So, I think one is an enabler of the other.

Abdiel Salas: Yes, time is always a constraint, so change needs to be managed.

Rick Rothermel: Yes—they probably have high employee engagement scores, because the organization leverages the key tenants of change management.

Michele Quinn: Absolutely.

Jessica Bronzert: Yes. Transformational change actually causes engagement to drop. Effective change management helps move employees through the disruption of change faster and with less disengagement to get back to a high engagement environment as quickly as possible.

Siobhan Pandya: Yes—employee engagement can change at any time, and therefore, change management should be a part of the organization on an ongoing basis. It can be incorporated into how we do communications, [the] process we follow for changes, etc.

16
Mini-Biographies of the Interviewees

Joe Beggiani: Director of Human Resources Operations, Global Talent & Learning Operations for Enterprise Services, Hewlett Packard Enterprise. Change master, Black Belt, program manager. Formerly with Bank of America and Motorola, Inc.

Darryl Bonadio: Director and Master Black Belt of the Office of Business Effectiveness Academy. MBA and Six Sigma certified since 2005. Profiled in the *Juran Quality Handbook* (6th edition). Over 35 years of experience across multiple industries and a quality improvement professional.

Dr. Gary Bradt: A keynote speaker on change and the author of *Change: The Tools You Need for the Life You Want at Work and Home*. Austin, TX: River Grove Books (December 2016).

Roxanne Britland: President of Professional Performance Group, LLC. Founder and former owner of International Society of Six Sigma Professionals (ISSSP). Currently, Organizational Development and Transformation consultant with the Federal Aviation Administration (FAA).

Jessica Bronzert: MBA, ACC, Principal at The Sparks Group and adjunct faculty at the Center for Creative Leadership. Formerly Director of Change Execution at Lowe's Companies, Inc.

Sarah Carleton: Master Black Belt, Principal of Sarah A. Carleton Consulting, providing Lean and Six Sigma training and mentoring. Author of *The Green Belt Memory Jogger* (GOAL/QPC, 2016).

Tom Cluley: Owner of Above the Fray Advisory Services, LLC. Formerly with The Wiremold Company. Author of several books, including *Driving Strategy to Execution Using Lean Six Sigma* (with Gerhard Plenert: CRC Press, 2012).

Currin Cooper: Over 20 years project and change management experience for Fortune 500 clients such as Bank of America, Wells Fargo, Coca-Cola Bottling Company. Consolidated, Lowe's Home Improvement, and Duke Energy. She is an adjunct instructor at the University of South Carolina Department of Continuing Education for PMI project management certification classes. She holds a PMP certification, a Six Sigma Green Belt, is a Certified Change Manager (CCM) through Acuity Institute, and a Certified Change Management Professional (CCMP) through AMCP.

Beth Cudney: Ph.D. ASQ Fellow, ASEM Fellow, Associate Professor, Director Design Engineering Center, Missouri University of Science and Technology, Formerly with Danaher and Dana Corporation. Author of several books, including *Using Hoshin Kanri to Improve the Value Stream* (CRC Press, 2009).

Bob Dodge: Co-Founder of IDI Consulting, and Former Director of Consulting Services for Lamarsh & Associates, Founder of Expert Change Management, and Sr. Vice President of The Alternative Board Denver West –Trusted Advisor to Business Owners as their Coach and Facilitator of Peer Advisory Boards.

Lynn Alvarez Doupsas: Principal, LD Consulting. Specific areas of expertise: program management, organizational change management, coaching, facilitation, employee engagement. Past clients: Verizon, Marriott, Disney / ABC Television, Mazda, Merrill Lynch. Speaker, writer, and regular contributor for many social networking, community, and special interest forums.

Mike Ensby: Manager with Ernst & Young, LLC, Performance Improvement Advisory. Specializing in change management issues facing project management office (PMO) structures and governance.

Marc Zaban Fogel: Principal Consultant at Dragon Thinking Consulting.

Scott Leek: Advisor, coach, and Master Black Belt at Sigma Consulting Resources, LLC. Author of several books and training courses, including contributing author of *Six Sigma Leadership Handbook* (John Wiley & Sons, 2003).

Don Linsenmann: CEO of Executive Transformation Mentoring, LLC. Previously, VP Business Process Excellence and Corporate Six Sigma Champion for DuPont. Coauthor of *The Six Sigma Fieldbook* with Mikel Harry.

Patra Madden: Advisory Engineer, Naval Nuclear Laboratories. Focusing on corporate transformation by utilizing Lean Six Sigma and change management tools and techniques. Author of absolutely no books.

Scott McAllister: Vice President of Growth at Prosci. Executive consultant; has worked with more than 80 clients in 28 countries to develop and deploy strategy, Lean Six Sigma, innovation, and change management initiatives.

Siobhan Pandya: Director of Continuous Improvement and Lean at Mary Kay, Inc. Formerly with Shell Oil Company in the UK and US. Lean Six Sigma Black Belt.

Michele Quinn: Certified Prosci Change Management Practitioner, Prosci Master Instructor and Train the Trainer professional, Former LSS Deployment Leader, and Master Black Belt.

David Ringel: Vice President of Operations at MainStream GS, LLC, where he leads the Performance Improvement and Change Management consulting business. Prior to joining MainStream in 2006, he was a successful operations executive in Tier 2 suppliers to the automotive industry.

Rick Rothermel: Owner of LaMarsh Global. Change management author and thought leader. Founding Board Member of the Association of Change Management Professionals.

Abdiel Salas: Master Black Belt and Lean Manager, Continuous Improvement Master, writer, blogger on spiritual issues, husband, and father of a successful family (best enterprise). Expertise: Worldwide implementer and mentor.

Bob Von Der Linn: President and Principal Consultant, Change Leadership Resources, LLC. Formerly with General Electric.

Section IV

Appendix A: Competencies for the Successful Black Belt

Competency area	Specific competency
Communication Skills	Strong, clear and candid communicator, and effective presenter with all levels within the organization, "quick on his/her feet," good negotiator, excellent writing skills.
	Listens effectively and stimulates the development of new ideas with effective probing and questioning.
Analytical, Technical, Project Management Skills	Demonstrates ability to dissect and comprehend complex and sometimes ill-defined situations.
	Willing to dive into details when necessary ... "get hands dirty."
Empowerment	Works effectively across traditional organizational boundaries of function, position, and other differences.
	Promotes cross-functional efforts. Trusts others. Encourages risk taking and empowerment behavior. Leverages individual strengths.
Passion and Enthusiasm	Creates exciting and positive working environment with strong projection of confidence for making improvements.
	Projects genuine personal commitment to business, quality, and process improvement. Passion for Lean Six Sigma.
Leadership, Training, Coaching	Demonstrates effective coaching skills. Motivates and effectively influences to ensure accomplishment of key objectives. Can lead without formal authority.
	Creates an open and receptive mindset for learning ... and effectively imparts new information.
Change Agent	Understands and promotes the need for change, open to ideas and change.
	Enjoys problem solving and initiates new and better ways of doing things.
Influential, "Makes Things Happen"	Speaks assertively and with authority of experience ... and is recognized as someone who will deliver results.
	Considered promotable to higher levels of responsibility due to consistent, excellent performance, high potential.
Confident	Doesn't hesitate to "push back" when he/she disagrees with direction.
	Willingness to face difficult issues ... take a stand ... become personally involved in resolving problems.

FIGURE A.1
Black Belt competency list. Some elements from The Lean Methods Group.

Appendix B: Coaching and Mentoring for the Black Belt

Sometimes, the Black Belt's boss is knowledgeable and experienced in the skill sets identified in the critical success factors triangle shown in Figure B.1. In that case, external coaching and mentoring relationships may not be required. But, in the case where the Black Belt works for a person who lacks training and experience in project management, Lean Six Sigma, and change management, it is recommended that a person (or persons) inside or outside the organization be found to serve as a coach and mentor for the Black Belt.

Some organizations establish these relationships with regard to the Lean Six Sigma skill set, but few provide a person to help with the issues that ultimately arise having to do with resistance and the mitigation thereof.

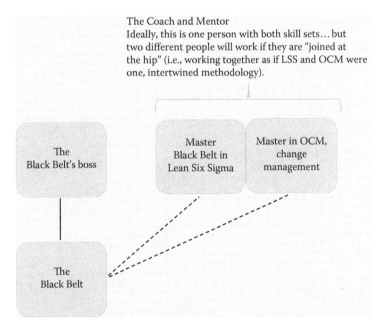

FIGURE B.1
Coaching and mentoring for the Black Belt.

Appendix C: The Shingo Principles

Reference *The Ten (10) Shingo Guiding Principles*—See http://www.singo.org/model.

The following are the first two principles:

1. Respect Every Individual
2. Lead with Humility

RESPECT EVERY INDIVIDUAL

Respect must become something that is deeply felt for and by every person in an organization. Respect for every individual naturally includes respect for customers, suppliers, the community, and the society in general. Individuals are energized when this type of respect is demonstrated. Most associates will say that to be respected is the most important thing they want from their employment. When people feel respected, they give far more than their hands—they give their minds and hearts as well.

To better understand the principle of respect for every individual simply ask the question, "why?" The answer is because we are all human beings with worth and potential. Because this is true, every individual deserves my respect.

EXAMPLES OF IDEAL "RESPECT-RELATED" BEHAVIORS

1. Create a development plan for employees that includes appropriate goals.
2. Involve employees in improving the work done in their areas.
3. Continually provide coaching for problem solving.

LEAD WITH HUMILITY

One trait common among leading practitioners of enterprise excellence is a sense of humility. Humility is an enabling principle that precedes learning and improvement. A leader's willingness to seek input, listen carefully, and continuously learn creates an environment where associates feel respected and energized and give freely of their creative abilities. Improvement is only possible when people are willing to acknowledge their vulnerability and abandon bias and prejudice in their pursuit of a better way.

EXAMPLES OF IDEAL, "HUMILITY-RELATED" BEHAVIORS

1. When an error occurs, focus on improving the process that created the error.
2. Ensure that all parts, materials, information, and resources are correct and meet specifications before using them in a process.

Appendix D: A Brief History of Continuous Improvement

Lean seems to be treated like a centuries-old religion by some. You may be surprised to learn that the term "Lean," as used in the continuous improvement context, was never uttered before 1990. As of the writing of this book, Lean is less than 40 years old!

- 1795—Carl Friedrich Gauss introduces concept of the normal curve
- 1913—Henry Ford introduces the assembly belt in his factories
- 1920s—Walter Shewhart's early work on process variation
- 1939—Edwards Deming publishes a series of Shewhart's lectures
- 1950s—Toyota Production System is developed by Kiichiro Toyoda, Taiichi Ohno, etc.
- 1950s—Shigeo Shingo introduces SMED and Poka-Yoke
- 1951—*Juran's Quality Handbook*, by Joseph M. Juran and A. Blanton Godfrey
- 1951—Kaizen is introduced in Japan
- 1960—Japan awards Dr. Deming the Order of the Sacred Treasure
- 1980s—Lean Manufacturing makes its way to the West. This decade marked the first use of the word "Lean" in the continuous improvement context
- 1986—Bill Smith at Motorola invents the Six Sigma quality improvement process
- 1990s—General Electric develops Six Sigma as a management approach
- 1990—*The Machine That Changed the World*, by James P. Womack, Daniel T. Jones, and Daniel Roos
- 1996—*Lean Thinking*, by James P. Womack and Daniel T. Jones
- 1998—*Lean Thinking*, by Mike Rother and John Shook

THE HISTORY OF CHANGE MANAGEMENT

Because this book is mostly about change management, I've included two perspectives on the history of this discipline from people who have been deeply involved in the niche for decades. First, an excerpt from an e-book written by Jeanenne LaMarsh, followed by an examination of the history and future of change management taken from the Prosci website—the organization founded by Jeff Hiatt.

A Brief History of Change Management

by Jeanenne LaMarsh

Do you make your living as a change management professional? Are you passionate about the intellectual challenges inherent during times of change and the positive difference you make in the lives of employees and leaders of the organizations you serve?

Then celebrate the fact that you are at a place where the need to manage changes well now intersects with real recognition by organizations that leaders must also deliver the resources to meet that need.

It wasn't always so.

Forty years ago, there was no change management. There were no structured, disciplined, nor transferable methodologies that could be internalized and leveraged by organizations. And certainly, there were no change management professionals.

Organizations have always handled changes, but until a few years ago, the few who did it well did it instinctually and often erratically. Unfortunately, most also did it from a frame of reference that did not take the fears, concerns, and questions of the people impacted into consideration. Randy Kesterson, VP of Global Ops, Op Excellence/TQM, Sourcing at Doosan, tells a story in his book *The Basics of Hoshin Kanri* of a related experience he had early in his career. While trying to convince a senior manager of the need to take the fears, concerns, and questions of those impacted into consideration before embarking on major changes, Randy tried to explain the use of a structured new concept called change management. With no hesitation, the senior leader held up his hand to signal Randy to stop talking. He said, "I already have a change management approach that has served me very well for over 30 years in business." The manager then swung his foot up on the conference table, pointed to it, and said "11-E," his shoe size, "If my people don't want to do it, I kick them all in the butt." That was his change management approach: the 11-E change management methodology.

Looking back on it, Randy's story is comical, but it shows the great progress we as change management advocates have made. History shows a

Appendix D: A Brief History of Continuous Improvement • 173

similar situation in the business process of managing inventory. Originally, there pretty much was no process. There was only "Joe" who knew where everything was in the warehouse or in the closet. "Go ask Joe. He'll find it for you…" was a simple inventory control system. This thinking became increasingly unsatisfactory as a way of managing the whereabouts of important business materials. And when World War II started, the explosion of need for equipment and materials to be shipped to Europe and Asia was too great to continue to rely on Joe with only his memory and his instincts. This led to the development of a structured organizational process for gaining control over the logistics of tracking, reordering, and shipping, which today is known as Inventory Control. Eventually, tools, processes, and systems were developed. Today, there is a profession and a set of professionals that are critical to organizations of all sizes all over the world directly related to the concept.

Another key effect of WWII was the development of what came to be called Management Science. Companies large and small had existed for centuries prior to the efforts of ten men, including Robert McNamara, who pioneered the field. These revolutionaries put structure and process to increases in operations, finance, and logistical information demanded by the size and scope of the war effort, and they basically created the Statistical Control profession.

And as the demands of the war resulted in the development of other new business disciplines, organizational innovation continued through the 1970s and 1980s with the need for companies to implement huge technical systems such as Manufacturing Resource Planning (MRP) and, subsequently Enterprise Resource Planning (ERP). Combined with the decision by management to introduce multiple major changes simultaneously, these new fields sparked the recognition by leaders that the need to help employees cope with all the changes happening around them was becoming increasingly critical.

This thinking triggered an idea in 1985. At the time, LaMarsh & Associates, which evolved into LaMarsh Global, had begun training on efficient ways for organizations to implement these important changes. We learned that the employees often took well to the new required behaviors and new technology solutions, but there were failures long term too often. And participants in workshops on team building, decision making, project planning and performance management helped demonstrate understanding of concepts and abilities to apply the tools and techniques used in the workshop setting, but failed to sustain the new learning.

Unfortunately many still failed to apply their newly acquired skills once they actually returned to the workplace.

This thinking led LaMarsh & Associates into the study of change and eventually change management. Our firm confirmed that training was really only one step in the process of changing behavior, and thus the organization. Our studies and analysis led to the development of a specific process and model for proactively managing crucial business changes. We named this deliberate approach to change management Managed Change.

THE HISTORY AND FUTURE OF CHANGE MANAGEMENT

Source: http://procsci.com/what-is-change-management

The Emergence of Our Discipline

Over the past quarter of a century, change management has emerged, evolved, and grown from foundational understandings to conceptual underpinnings and on to a recognized discipline. Prosci's research and experience suggest that in the coming years the focus will shift toward advancement along three fronts:

1. Increased collaboration of change disciplines
2. Enhanced organizational maturity development
3. Individual professional development

Brief History of Change Management

Four distinct eras mark the evolution and growth of the change management discipline:

1. *Pre-1990s: Foundations.* Academics begin to understand how humans and human systems experience change.
2. *1990s: On the radar.* Change management enters the business vernacular.
3. *2000s: Formalization.* Additional structure and rigor codify change management as a discipline.
4. *Going forward.* Individual professional development and enhanced growth of organizational maturity emerge.

Appendix D: A Brief History of Continuous Improvement • 175

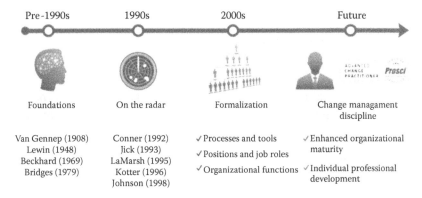

FIGURE D.1
Change management timeline.

Pre-1990: Foundations

The first era of change management was the period before 1990. During this period, the focus was on improving our collective understanding of human beings, how we experience change, and how our human systems interact and react. This era provided crucial insights, research, and frameworks for understanding successful change. Some of the primary contributors during this time include:

Arnold van Gennep (1909)

Van Gennep was a cultural anthropologist studying rites of passage around the globe. He introduced the idea of change as happening in three states: separating from our current state, moving through a transition, and reincorporating into a future state (examples include adolescence, marriage, and parenthood).

Kurt Lewin (1948)

A social psychologist, Lewin introduced three states of change—unfreezing, moving, and refreezing—as well as force field analysis.

Richard Beckhard (1969)

Beckhard was a pioneer in organization development and defined the discipline as "an effort (1) planned, (2) organization-wide, and (3) managed from the top, to (4) increase organization effectiveness and health through (5) planned interventions in the organization's 'processes,' using behavioral-science knowledge."

William Bridges (1979)

Speaker, author, and consultant, Bridges described the states of a transition as the ending, the neutral zone, and the new beginning.

This list is certainly not exhaustive. Many other scholars from psychology, business, engineering, and the social sciences added research and insights that helped shape our understanding of how people experience change. It is important to understand the roots.

> *Contribution of the Foundations era: An underlying understanding of how individuals and systems experience change.*

1990–2000: On the RADAR

The second era of change management was the decade of the 1990s. During the "on the radar" era, change management began to enter the business vernacular. The people side of change moved out of the academic and exploratory space and into concepts discussed at project meetings, in C-suites and around boardroom tables. Language began to form around the discipline of change management, and many of the guiding principles that still guide the discipline were articulated during this time. The first steps were taken to show that individual change does not happen by chance, but can be supported and driven with thoughtful and repeatable steps.

While there were many who contributed to putting change management on the radar, some of the most notable include:

General Electric (Early 1990s)

As a recognized world leader, GE introduced the Change Acceleration Process as part of its larger improvement program.

Daryl Conner (1992)

In his seminal work, *Managing at the Speed of Change* Conner provided invaluable insights on numerous change concepts and topics.

Todd Jick (1993)

Jick's *Managing Change: Cases and Concepts* included topical case studies and a chapter titled "Implementing Change" in which he shed light on the common pitfalls and introduced his Ten Commandments of Implementing Change.

Jeanenne LaMarsh (1995)

LaMarsh's work *Changing the Way We Change* developed concepts around the importance of the ability to change, the mitigation of resistance, and the enabling frameworks for supporting change.

John Kotter (1996)

First in a Harvard Business Review article and later in his book, *Leading Change*, Kotter described eight change failure modes and subsequent steps to address them.

Spencer Johnson (1998)

Johnson's *Who Moved my Cheese?* presents readers with a parable that addresses how people can deal with the changes happening around them and to them.

During the 1990s, change management landed "on the radar" with concepts and language that began to take hold in mainstream management. Geopolitical forces, economic development, budding new value systems (think empowerment), and shifting employee–employer relationships set the stage for an increased recognition of how important the human side of change was.

Contribution of the On the Radar era: Socialization of the phraseology and importance of the people side of change and conceptual underpinnings of an emerging discipline.

2000–Present: Formalization

The third era in the development of change management was that of the 2000s, leading up to the present. This era of change management was marked by the formalization of the discipline. Where the foundations era gave us underlying understanding, and the "on the radar" era gave us concepts and language, a shift occurred as we entered the new millennium.

Growing out of a need for greater repeatability and structure, the change management discipline began adding formal structure and discipline on a number of fronts. Although founded in 1994, it was early in the 2000s that Prosci formalized and accelerated its research specifically in change management.

In 2003, Prosci introduced the first integrated approach to change management that leveraged organizational and individual change management processes and tools. The Prosci ADKAR Model, an individual change model, provided an outcome orientation to change management work: driving success one person at a time. The Prosci 3-Phase Process provided the structure, process, and tools for creating customized change management strategies and plans. In 2005, Prosci started to formalize research and a platform for the innovators of the discipline, who started working to embed change management as a core capability of their organizations.

There are three dimensions where this formalization can be seen more than any others:

Processes and Tools

Building on the underlying understandings and concepts that had been laid, practitioners began building and applying more rigorous structure to the work of change management, including more robust and repeatable processes and enhanced tools to support consistent application.

Positions and Job Roles

Organizations began creating specific jobs with a sole focus on applying change management on projects and initiatives (with significant growth since 2010).

Organizational Functions

Organizations began establishing and resourcing functions and structures to support change management application across the enterprise (such as a Change Management Office, Center of Excellence, or Community of Practice).

In addition to formalizing of processes and tools, positions and job roles, and organizational functions, steps were taken to formalize the profession of change management during this era. Professional associations, standards, and certifications emerged during this time (such as the Change Management Institute and the Association of Change Management Professionals).

> *Contribution of the Formalization era: Structure and definition, along with tools and processes, that provided repeatability and credibility to a growing discipline.*

Appendix E: Where to Go for More Information

Organizational Change Management
Tool: ADKAR Model (and others)
Organization: Prosci
Website: www.prosci.com

Tool: Managed Change Model
Organization: lamarsh Global
Website: www.lamarsh.com

Organization: Conner Partners
Website: www.connerpartners.com

Tool: General Electric CAP (Change Acceleration Program) Model
Website: https://www.slideshare.net/HomerZhang/
 ge-change-managementcap

Lean Six Sigma (Sources of Free Information)
Organization: Lean Methods Group (formerly Breakthrough Management Group)
Website: www.leanmethods.com

Organization: iSixSigma
Website: https://www.isixsigma.com/

People Are Different
Tool: DISC Profile
Source: Multiple sources and websites identified

Tool: WorkPlace Big Five Profile
Organization: Center for Applied Cognitive Studies

Website: http://www.CENTACS.com (soon to be http://www.ParadigmPersonality.com)

Tool: Change Style Indicator
Publisher: Multi-Health Systems Inc.
Website: https://ecom.mhs.com

Appendix F: Acknowledgments

What follows is a list of the people who helped me with the editing of the final draft of this book. If I have accidentally excluded your name, I apologize.

I am thankful to those who allowed me to interview you for this book: Joe Beggiani, Darryl Bonadio, Dr. Gary Bradt, Roxanne Britland, Jessica Bronzert, Sarah Carleton, Tom Cluley, Currin Cooper, Beth Cudney, PhD, Bob Dodge, Lynn Alvarez Doupsas, Mike Ensby, Marc Fogel, Scott Leek, Don Linsenmann, Patra Madden, Scott McAllister, Siobhan Pandya, Michele Quinn, David Ringel, Rick Rothermel, Abdiel Salas, Bob Von Der Linn.

Also to David Silverstein for writing the Foreword; to Pierce J. Howard, PhD and Zack Johnson for writing a special note for use in the book; to Siobhan Pandya and Michele Quinn, and Alireza (Ali) Kar for allowing me to use your quotes; to Ellen Domb, PhD for your candid input and help with editing; to the fantastic team at Productivity Press to include Michael Sinocchi, Alexandria Gryder, and Mathi Ganesan; and to the following for providing endorsements.

Kind words about this book from Marshall Goldsmith, Ellen Domb, PhD and Gerhard Plenert, PhD can be found on the back cover. Other endorsements, including those from the people listed below, are available on my website at RandyKesterson.com.

Jon Atwood, Janet Basile, Christopher Biggs, James Bond, Andy Bordick, Tom Bornemann, Kevin Briede, Jessica Bronzert, Paul Carney, Joyce Carroll, Louis Carter, Jason T. Collett, Currin Cooper, Bob Dodge, Ellen Domb, PhD, Erdem Dursun, Mark Evans, David Foxx, Marshall Goldsmith, Brent Grazman, PhD, Dr. Steve Griffin, Paul Grizzell, Jamie "Jay" Guttenberg, PhD (ABD), Stephen G. Hall, Robert Handfield, PhD, Erik T. Hansen, Arun Hariharan, Jesse Harrington, Juan Victor Hernandez, Owen Hewitt, Steven Hodlin, Pierce J. Howard, PhD, Anoop Jain, Zack Johnson, Mark Jolley, Robert Jones, Simranjit Kainth, Gary Kapanowski, Koray Karakas, Kay Kendall, Christine Kreuser, Jeanenne LaMarsh, Don Linsenmann, Craig Long, Andrew S. McCune, Rear

Admiral Terry McKnight, USN (Retired), Kevin McManus, Denise Meredith, Cornelius Moore, Richard Moormann, Terry Newell, Zane R Nobbs, Patrick O'Hearn, Siobhan Pandya, Dave Parkerson, Ralph W."Pete" Peters, Gerhard Plenert, PhD, James Michael Reames, Karen D. Riding, PhD, Teresa Riffel, Jorge J. Roman, Neal Ropski, Lori F, Rothenberg, PhD, Rick Rothermel, Paul Sample, Amy Sequeira, Jeremy A. Sparks, Dean R. Spitzer, PhD, Mark Stewart, Barbara A. Trautlein, PhD, Frank Uhelsky, John Vaughn, Albert Vermeulen, Sandra West, Matthew Wilson, Catherine Wolpert, Chris Woodbine, Doug Wright, Herman Zwirn.

Bibliography

J. M. Hiatt and T. J. Creasey, *Change Management: The People Side of Change*, Prosci Learning Center Publications, Loveland, CO 2003.

R. Potts and J. LaMarsh, *Master Change, Master Success*, Chronicle Books, San Francisco, CA 2004.

Index

A

Accommodation (A), 80
ADKAR Model, 37, 42
American simplified keyboard, 17
Aristotle, 12
Arm-crossing example, 21, 22
Awareness and desire, 31–33, 37

B

Balanced Scorecard, 26, 27, 28
The Basics of Hoshin Kanri (book), 30, 172
Beckhard, Richard, 175
BeeRaider keyboard, 17, 18
Beggiani, Joe (interview with)
    addressing group of thinkers and feelers, 152
    biography of, 159
    change management
        BBs asking for help, 140
        biggest mistake, with regard to employing, 135
        describing, 106
        dos and don'ts, 126
        effectiveness, measurement of, 132
        employee engagement and, 156
        failure/risk, 122
        methodology, 130
        own words about, 103
        personal experience with, 93
        project champion's role, 145–146
        reasons for using, 107
    gaining adoption and eliminating resistance, 149
    Lean Six Sigma, personal experience with, 97
    measuring resistance, 142
    OCM and LSS intersection
        critical success factors, 116
        explaining the importance of, 99
        lessons learned, 119
        needs, 138
        success, 109
        taught at BB training, 113
Behavior change, tracking, 34
Belasco, James, 19
Black Belt
    coaching and mentoring for, 167
    competency list, 165
    critical success factors for, 4–5
    successful, 75
Bonadio, Darryl (interview with)
    addressing group of thinkers and feelers, 153
    biography of, 159
    change management
        BBs asking for help, 140
        biggest mistake, with regard to employing, 136
        describing, 106
        dos and don'ts, 126–127
        effectiveness, measurement of, 133
        employee engagement and, 156
        failure/risk, 123
        methodology, 130
        own words about, 104
        personal experience with, 94
        project champion's role, 146
        reasons for using, 107–108
    gaining adoption and eliminating resistance, 150
    Lean Six Sigma, personal experience with, 97
    measuring resistance, 142–143
    OCM and LSS intersection
        critical success factors, 118
        explaining the importance of, 100–101
        lessons learned, 120
        needs, 138
        success, 110
        taught at BB training, 114
Bonaparte, Napoleon, 12

Bradt, Gary (interview with)
	addressing group of thinkers and feelers, 152
	biography of, 159
	change management
		BBs asking for help, 140
		biggest mistake, with regard to employing, 135–136
		describing, 106
		dos and don'ts, 126
		employee engagement and, 156
		failure/risk, 123
		methodology, 130
		own words about, 103–104
		personal experience with, 94
		project champion's role, 146
		reasons for using, 107
	gaining adoption and eliminating resistance, 149
	measuring resistance, 142
	OCM and LSS intersection
		critical success factors, 117
		explaining the importance of, 100
Branson, Richard, 13
Bridges, William, 176
Britland, Roxanne (interview with)
	addressing group of thinkers and feelers, 153
	biography of, 159
	change management
		BBs asking for help, 140
		biggest mistake, with regard to employing, 136
		describing, 106
		dos and don'ts, 126
		effectiveness, measurement of, 133
		employee engagement and, 156
		failure/risk, 123
		methodology, 130
		own words about, 104
		personal experience with, 94
		project champion's role, 146
		reasons for using, 107
	Lean Six Sigma, personal experience with, 97
	measuring resistance, 142
	OCM and LSS intersection
		critical success factors, 117
		explaining the importance of, 100
		lessons learned, 120
		needs, 138
		success, 110
		taught at BB training, 113
Bronzert, Jessica (interview with)
	addressing group of thinkers and feelers, 155
	biography of, 159
	change management
		BBs asking for help, 141
		biggest mistake, with regard to employing, 137
		describing, 107
		effectiveness, measurement of, 134
		employee engagement and, 157
		failure/risk, 124–125
		methodology, 131
		own words about, 106
		personal experience with, 96
		project champion's role, 148
		reasons for using, 108
	gaining adoption and eliminating resistance, 151
	OCM and LSS intersection
		explaining the importance of, 103
		lessons learned, 121–122
		needs, 139
		taught at BB training, 115

C

Carleton, Sarah (interview with)
	addressing group of thinkers and feelers, 152
	biography of, 159
	change management
		BBs asking for help, 140
		biggest mistake, with regard to employing, 136
		describing, 106
		dos and don'ts, 126
		effectiveness, measurement of, 132
		employee engagement and, 156
		failure/risk, 123
		own words about, 104
		personal experience with, 94
		project champion's role, 146
		reasons for using, 107

Index

gaining adoption and eliminating resistance, 149
Lean Six Sigma, personal experience with, 97
measuring resistance, 142
OCM and LSS intersection
  critical success factors, 117
  explaining the importance of, 100
  lessons learned, 120
  needs, 138
  success, 109
  taught at BB training, 113
"Catch ball" process, 30
Center for Applied Cognitive Studies (CENTACS)
  WorkPlace Big Five Profile, 78–82
Champion, 55, 57
Change, 11–13, 15, 19–20, 21, *see also* Organizational change management (OCM)
  bad changes, 57–58
  chaordic, 47–48
  five stages of, 46
  people, to deal with, 77–84
  prevention of bad projects, 58
  reasons for, 31–32
  states of, 175
  targets of, 32–33, 60
Change acceleration process (CAP), 28, 176
Change curve, 21–23
  with and without well-managed change, 44–45
Change management, *see also* Organizational change management (OCM)
  1990–2000, 176–177
  2000–present, 177–178
  history of, 172–177
  pre-1990, 175–176
  timeline, 175
Change manager, 81
Change propensity, 82–83
Change readiness checklist, 55, 56
"Change receptiveness" of organization, 6–7
Change Style Indicator, 80–81
Changing the Way We Change (book), 177
Chaordic change, 47–48

Churchill, Winston, 12
Cluley, Tom (interview with)
  addressing group of thinkers and feelers, 151
  biography of, 159
  change management
    BBs asking for help, 139
    biggest mistake, with regard to employing, 135
    describing, 106
    dos and don'ts, 125
    effectiveness, measurement of, 132
    employee engagement and, 155
    failure/risk, 122
    methodology, 129
    own words about, 103
    personal experience with, 93
    project champion's role, 145
    reasons for using, 107
  gaining adoption and eliminating resistance, 148
  Lean Six Sigma, personal experience with, 96
  measuring resistance, 141
  OCM and LSS intersection
    critical success factors, 116
    explaining the importance of, 99
    lessons learned, 119
    needs, 137
    success, 109
    taught at BB training, 112
Collaboration, 8
Communication, 66–67
Communication plan, 63–65
Completion, 46
Conner, Daryl, 45–46, 176
Consolidation (C), 80
Continuous improvement, 8, 51, 52
  brief history of, 171–178
Cook, Claire, 13
Cooper, Currin (interview with)
  addressing group of thinkers and feelers, 153–154
  biography of, 160
  change management
    BBs asking for help, 140
    describing, 106
    dos and don'ts, 127

Cooper, Currin (interview with) (*cont.*)
   effectiveness, measurement of, 133–134
   employee engagement and, 156–157
   failure/risk, 124
   methodology, 130–131
   own words about, 104–105
   personal experience with, 95
   project champion's role, 147
   reasons for using, 108
  gaining adoption and eliminating resistance, 150
  Lean Six Sigma, personal experience with, 98
  measuring resistance, 143
  OCM and LSS intersection
   explaining the importance of, 101
   lessons learned, 120
   needs, 139
   taught at BB training, 114
Creasey, Tim, 42
Critical success factors, 4
Crossed arms experiment, 21, 22
Crystal, Billy, 20
C's (communication, cooperation, collaboration), 7, 8, 70
Cudney, Beth (interview with)
  addressing group of thinkers and feelers, 153
  biography of, 160
  change management
   BBs asking for help, 140
   biggest mistake, with regard to employing, 136
   describing, 106
   dos and don'ts, 127
   effectiveness, measurement of, 133
   employee engagement and, 156
   failure/risk, 123
   own words about, 104
   personal experience with, 95
   project champion's role, 147
   reasons for using, 108
  gaining adoption and eliminating resistance, 150
  Lean Six Sigma, personal experience with, 97–98
  measuring resistance, 143
  OCM and LSS intersection
   critical success factors, 118
   explaining the importance of, 101
   lessons learned, 120
   needs, 139
   success, 111
   taught at BB training, 114
Current State model, LaMarsh's, 43

D

Darwin, Charles, 13
da Vinci, Leonardo, 20
Dealey, William, 17
Decay, 11
Delta State, 46–47
Deming, E., 12, 171
Desire and awareness, 31–33, 37
Desired Future State, 43–44
DISC model, 77–78
DMAIC process, 73–74
Dodge, Bob (interview with)
  addressing group of thinkers and feelers, 151
  biography of, 160
  change management
   BBs asking for help, 139–140
   biggest mistake, with regard to employing, 135
   describing, 106
   dos and don'ts, 125
   effectiveness, measurement of, 132
   employee engagement and, 155
   failure/risk, 122
   methodology, 129
   own words about, 103
   personal experience with, 93
   project champion's role, 145
   reasons for using, 107
  gaining adoption and eliminating resistance, 148
  Lean Six Sigma, personal experience with, 96
  measuring resistance, 141
  OCM and LSS intersection
   critical success factors, 116
   explaining the importance of, 99
   lessons learned, 119
   needs, 137
   success, 109

Index • 189

taught at BB training, 112
Domb, Ellen, 9, 88
Don, Pastor, 48
Doupsas, Lynn (interview with)
   addressing group of thinkers and feelers, 152
   biography of, 160
   change management
      biggest mistake, with regard to employing, 135
      describing, 106
      effectiveness, measurement of, 132
      employee engagement and, 156
      methodology, 130
      own words about, 103
      personal experience with, 93
      project champion's role, 146
      reasons for using, 107
   Lean Six Sigma, personal experience with, 97
   measuring resistance, 142
   OCM and LSS intersection
      critical success factors, 116
Dvorak, August, 17
Dvorak keyboard, 17

E

"11-E approach to change management," 29
Emotional cycle of change model, 22, 45–46
Emotional response/performance, 23
Emotional Response to Change, 22
Emotional support, 47–49
Ensby, Mike (interview with)
   addressing group of thinkers and feelers, 154
   biography of, 160
   change management
      BBs asking for help, 140
      biggest mistake, with regard to employing, 136
      describing, 106
      effectiveness, measurement of, 134
      employee engagement and, 157
      failure/risk, 124
      methodology, 131
      own words about, 105
      personal experience with, 95
      project champion's role, 147
      reasons for using, 108
   gaining adoption and eliminating resistance, 150
   Lean Six Sigma, personal experience with, 98
   measuring resistance, 143
   OCM and LSS intersection
      critical success factors, 118
      explaining the importance of, 101–102
      lessons learned, 120–121
      needs, 139
      success, 111
      taught at BB training, 114–115
Extraversion (E), 79

F

Facts and data, 82–83
The Five Stages of a Change, 22
Five Stages of Grief, 22, 45–47
Fogel, Marc Zaban (interview with)
   addressing group of thinkers and feelers, 153
   biography of, 160
   change management
      BBs asking for help, 140
      biggest mistake, with regard to employing, 136
      describing, 106
      dos and don'ts, 127
      methodology, 130–131
      own words about, 104
      personal experience with, 95
      project champion's role, 147
   gaining adoption and eliminating resistance, 150
   Lean Six Sigma, personal experience with, 97
   measuring resistance, 143
   OCM and LSS intersection
      critical success factors, 118
      explaining the importance of, 101
      lessons learned, 120
      needs, 138
      success, 111
      taught at BB training, 114

Ford, Henry, 171
Franklin, Benjamin, 13
Friction equation, 49–50
Future State model, LaMarsh's, 43–44

G

Gauss, Carl Friedrich, 171
General Electric, 28, 171, 176
Gennep, Arnold Van, 175
Godfrey, A. Blanton, 73
Grief Cycle, 22

H

Hiatt, Jeff, 31, 42, 172
Hock, Dee, 48
Hopeful realism, 46
Howard, Pierce J., 81
Humility, 170
    respect and, 7, 70, 75

I

Imperial System, 20
Incentives, 66–67
Influencers, 32–33
Informed optimism, 46
Informed pessimism, 46
Interviews with Change Management experts, 91–158
Intuition, about change management, 70–71
Inventory Control, 173

J

Jick, Todd, 176
Johnson, Spencer, 177
Johnson, Zack, 35
Juran's Quality Handbook, 171

K

Kaizen, 171
KALQ keyboard, 17, 18
Kelley, Don, 45–46
Kennedy, Robert, 19
Kettering, Charles, 13

Keyboard ball, 18, 19
Kirkpatrick Model, 67–68
Kotter, John, 31, 177
Kübler-Ross, Elisabeth, 45
Kübler-Ross model, 22, 45–47

L

LaMarsh, Jeanenne, 31, 46, 60, 172, 177
LaMarsh Managed Change methodology, 42–45
Leadership, 8
Leading Change (book), 31, 177
Lean (term), 171
Lean Six Sigma (LSS), 3, 25, 26, 27, 28
    interviews with experts about, 93–158
    training programs, 73
        deployment, 75
        intersection of OCM and, 74–75
        tool box with, 75
    websites, information on, 179
Lean Thinking (book), 171
Lean Transformation, 69
Leek, Scott (interview with)
    addressing group of thinkers and feelers, 152
    biography of, 160
    change management
        BBs asking for help, 140
        biggest mistake, with regard to employing, 135
        describing, 106
        effectiveness, measurement of, 132
        employee engagement and, 156
        failure/risk, 123
        personal experience with, 93–94
        project champion's role, 146
    gaining adoption and eliminating resistance, 149
    Lean Six Sigma, personal experience with, 97
    measuring resistance, 142
    OCM and LSS intersection
        critical success factors, 116–117
        explaining the importance of, 99–100
        lessons learned, 119–120

needs, 138
success, 109
taught at BB training, 113
Lewin, Kurt, 175
Lichtenberg, Georg C., 12
Linn, Bob Von Der (interview with)
  addressing group of thinkers and feelers, 153
  biography of, 161
  change management
    BBs asking for help, 140
    biggest mistake, with regard to employing, 136
    describing, 106
    effectiveness, measurement of, 133
    employee engagement and, 156
    failure/risk, 123
    methodology, 130
    own words about, 104
    personal experience with, 94
    project champion's role, 146–147
    reasons for using, 108
  gaining adoption and eliminating resistance, 150
  Lean Six Sigma, personal experience with, 97
  measuring resistance, 143
  OCM and LSS intersection
    critical success factors, 118
    explaining the importance of, 101
    lessons learned, 120
    needs, 138
    success, 110–111
    taught at BB training, 114
Linsenmann, Don (interview with)
  biography of, 160
  change management
    describing, 106
    own words about, 104
    personal experience with, 94
    reasons for using, 108
  Lean Six Sigma, personal experience with, 97
  OCM and LSS intersection
    critical success factors, 118
    explaining the importance of, 101
    success, 110
    taught at BB training, 114

M

Machiavelli, Niccolo, 19
The Machine That Changed the World (book), 171
Madden, Patra (interview with)
  addressing group of thinkers and feelers, 154
  biography of, 161
  change management
    BBs asking for help, 141
    biggest mistake, with regard to employing, 136
    describing, 106
    dos and don'ts, 127
    effectiveness, measurement of, 134
    employee engagement and, 157
    failure/risk, 124
    methodology, 131
    own words about, 105
    personal experience with, 95
    project champion's role, 147
    reasons for using, 108
  gaining adoption and eliminating resistance, 150
  Lean Six Sigma, personal experience with, 98
  measuring resistance, 143–144
  OCM and LSS intersection
    critical success factors, 118
    explaining the importance of, 102
    lessons learned, 121
    needs, 139
    success, 111
    taught at BB training, 115
Managed Change model, 42–45
Management Science, 173
Managing Change: Cases and Concepts (book), 176
McAllister, Scott (interview with)
  addressing group of thinkers and feelers, 154
  biography of, 161
  change management
    BBs asking for help, 141
    biggest mistake, with regard to employing, 137
    describing, 106
    dos and don'ts, 127–128

McAllister, Scott (interview with) (cont.)
    effectiveness, measurement of, 134
    employee engagement and, 157
    failure/risk, 124
    methodology, 131
    own words about, 105
    personal experience with, 95
    project champion's role, 147
    reasons for using, 108
  gaining adoption and eliminating resistance, 150
  Lean Six Sigma, personal experience with, 98
  measuring resistance, 144
  OCM and LSS intersection
    critical success factors, 118
    explaining the importance of, 102
    lessons learned, 121
    needs, 139
    success, 111
    taught at BB training, 115
McEnaney, Ray, 17
McNamara, Robert, 173
Metric System, 20
Multi-Health Systems Inc., 81

N

Need for stability (N), 79, 80
Negative change, 22
On Death and Dying (book), 45

O

Openness, 80
Operational excellence, 7
Order of the Sacred Treasure award, 171
Organizational change management (OCM), 3, 28, 29
  basic tools, 59–68
  communication plan, 63–65
  component systems (technical/administrative/people), 81
  definition of, 41
  emotional support, 47–49
  five stages, 45–47
  friction equation, 49–50
  intersection of Lean Six Sigma (LSS) and, 74–75
  interviews with experts about, 93–158
  intuition about, 70–71
  Lean/Continuous Improvement initiative, 50–52
  "lever pulling," 65–66
  Managed Change, 42–45
  Prosci ADKAR Model for, 37, 42
  risk analysis tool, 62–63
  seek out help, in dealing with resistance, 69–70
  stakeholder analysis tool, 59–61
  tool box with, 75
  tools mapping to DMAIC steps, 74–75
  websites, information on, 179
Originality (O), 79

P

Pandya, Siobhan (interview with)
  addressing group of thinkers and feelers, 155
  biography of, 161
  change management
    BBs asking for help, 141
    biggest mistake, with regard to employing, 137
    describing, 107
    dos and don'ts, 129
    effectiveness, measurement of, 134
    employee engagement and, 157
    failure/risk, 125
    methodology, 131
    own words about, 105–106
    personal experience with, 96
    project champion's role, 148
    reasons for using, 109
  Lean Six Sigma, personal experience with, 98–99
  measuring resistance, 145
  OCM and LSS intersection
    critical success factors, 119
    explaining the importance of, 103
    lessons learned, 121
    needs, 139
    success, 111
    taught at BB training, 115
Pandya, Siobhan, 3
People
  dealing with change, 77–84

and facts and data, 82–83
workplace personalities and thinking styles, 78–82
People Analytics, 83–84
People analytics tools, 33–35
People's workplace personalities, 78–82
Project charter risk assessment, 57
Project charter template, 55, 56
Project management, 26, 27
 tool box with, 75
Project risk assessment, 55–58
Prosci ADKAR Model, 37, 42, 178

Q

QualityGurus.com, 73
Questions, interview, 92–93
Quinn, Michele (interview with), 41
 addressing group of thinkers and feelers, 155
 biography of, 161
 change management
  BBs asking for help, 141
  biggest mistake, with regard to employing, 137
  describing, 106
  dos and don'ts, 129
  effectiveness, measurement of, 134
  employee engagement and, 157
  failure/risk, 124
  methodology, 131
  own words about, 105
  personal experience with, 96
  project champion's role, 148
  reasons for using, 108
 gaining adoption and eliminating resistance, 151
 Lean Six Sigma, personal experience with, 98
 measuring resistance, 144
 OCM and LSS intersection
  critical success factors, 119
  explaining the importance of, 102–103
  lessons learned, 121
  needs, 139
  success, 111
  taught at BB training, 115
QWERTY keyboard, 15–17, 18

R

Reinforcements, 66–67
Resistance, 21–23
 to change plan, 44–45
 to leaving Current State, 43
 minimizing, 25–30
 to moving to Future State, 43–44
 in personal life, 37
Respect
 for every individual, 169
 and humility, 7, 70, 75
Riegel, Lisa, 13, 67
Ringel, David (interview with)
 addressing group of thinkers and feelers, 152
 biography of, 161
 change management
  BBs asking for help, 140
  biggest mistake, with regard to employing, 135
  describing, 106
  effectiveness, measurement of, 132
  employee engagement and, 155
  failure/risk, 122
  methodology, 130
  own words about, 103
  personal experience with, 93
  project champion's role, 145–146
  reasons for using, 107
 gaining adoption and eliminating resistance, 149
 Lean Six Sigma, personal experience with, 96
 measuring resistance, 142
 OCM and LSS intersection
  critical success factors, 116
  explaining the importance of, 99
  lessons learned, 119
  needs, 138
  success, 109
  taught at BB training, 113
Risk analysis tool, 62–63
Rothermel, Rick (interview with)
 addressing group of thinkers and feelers, 154
 biography of, 161
 change management
  BBs asking for help, 141

Rothermel, Rick (interview with) (cont.)
   biggest mistake, with regard to employing, 137
   describing, 106
   dos and don'ts, 128
   effectiveness, measurement of, 134
   employee engagement and, 157
   failure/risk, 124
   methodology, 131
   own words about, 105
   personal experience with, 96
   project champion's role, 147–148
   reasons for using, 108
  gaining adoption and eliminating resistance, 151
  Lean Six Sigma, personal experience with, 98
  measuring resistance, 144
  OCM and LSS intersection
   critical success factors, 118–119
   explaining the importance of, 102
   lessons learned, 121
   needs, 139
   taught at BB training, 115

S

Salas, Abdiel (interview with)
  addressing group of thinkers and feelers, 154
  biography of, 161
  change management
   BBs asking for help, 141
   biggest mistake, with regard to employing, 136
   describing, 106
   dos and don'ts, 128
   effectiveness, measurement of, 134
   employee engagement and, 157
   own words about, 105
   personal experience with, 95
   project champion's role, 147
   reasons for using, 108
  gaining adoption and eliminating resistance, 150
  Lean Six Sigma, personal experience with, 98
  measuring resistance, 144

OCM and LSS intersection
  critical success factors, 118
  explaining the importance of, 102
  lessons learned, 121
  success, 111
  taught at BB training, 115
Senge, Peter, 20
Sharma, Robin, 20
Shewhart, Walter, 171
Shingo principles, 169–170
Shingo Prize for Excellence (award), 7
Shingo, Shigeo, 7, 171
Sholes, Christopher, 15
Silverstein, David, 8
Simplified keyboard, 17
Sinek, Simon, 12
Smith, Bill, 171
Spherical keyboards, 18, 19
Sponsor, 55
Stakeholder analysis tool, 59–61
Stakeholders, 59
Stamp, Jimmy, 18
Stayer, Ralph, 19
Stevenson, Adlai E., 12
Strategy formulation and deployment approach
  version 1.0, 25–26
  version 1.1, 26–27
Success
  catalyst for, 8
  critical factors, 4
Swift, Taylor, 91
Syndio Social, 35

T

Targets, of the change, 32–33, 60
Team of Teams: New Rules of Engagement for a Complex World (book), 8–9
Tolstoy, Leo, 19
Toyota Production System, 7, 171
Training, 66
  and development, 67

U

Uninformed optimism, 46
Utah State University, 7

V

Valley of Despair, 23, 46
von Goethe, Johann Wolfgang, 12

W

WASTE, 9, 88
Welch, Jack, 12
Wells, H.G., 12
Who Moved My Cheese? (book), 177
WIIFM, 31–32, 48
Wilson, Woodrow, 19
WorkPlace Big Five Profile, 78–82